Exploring the Universe of
Opportunities in the Space Industries

COSMIC
CAREERS

ALASTAIR STORM BROWNE
and MARYANN KARINCH

HarperCollins
Leadership

An Imprint of HarperCollins

Published by HarperCollins Leadership, an imprint of HarperCollins Focus LLC.

Any internet addresses, phone numbers, or company or product information printed in this book are offered as a resource and are not intended in any way to be or to imply an endorsement by HarperCollins Leadership, nor does HarperCollins Leadership vouch for the existence, content, or services of these sites, phone numbers, companies, or products beyond the life of this book.

ISBN 978-1-4002-2085-4 (eBook)
ISBN 978-1-4002-2093-9 (PBK)

Library of Congress Control Number: 2020948936

Printed in the United States of America
20 21 22 23 LSC 10 9 8 7 6 5 4 3 2 1

To Robert Kits van Heyningen, childhood friend, inventor, and co-founder of KVH Industries. He showed us how far one can go when he or she uses their full potential. An example for all of us to follow. —Alastair

To Jim McCormick, my daily source of information and inspiration. —Maryann

CONTENTS

PART III—The Technical Infrastructure

PART IV—The Adventure

INTRODUCTION

Past, Present, and Back to the Future

In October 1947, *Glamorous Glennis* zoomed through the skies with Captain Charles "Chuck" Yeager piloting. The flight of this rocket plane celebrated the accomplishments of our first modern space workers: scientists, engineers, technicians, administrators, pilots, and "ordinary" laborers. They were ordinary in the sense that their skills with tools and machines could have been used in any number of manufacturing plants. They were extraordinary in the sense that the quality of their work had dramatic consequences: triumph or tragedy, life or death.

The greatest achievement in any space program was no doubt the Apollo Moon landings. It was as though all of humanity jumped into a time machine and got a glimpse of the future. From President

Kennedy's challenge in 1961 to the first lunar landing in 1969, it took the dedication of thousands of scientists, engineers, technicians, administrators, pilots, and extraordinary laborers to accomplish this feat. More than 300,000 technicians alone contributed to the eight-year effort to put Neil Armstrong and Buzz Aldrin on the surface of the Moon.[1]

The legacy of the Moon landing is not a single technical achievement, however. NASA and the astronauts inspired generations of humans to pursue careers that have transformed and upgraded our lives. They spun off technologies from the Apollo Moon project and began myriad new ventures, giving us an even stronger sense of momentum forward in time. From their innovations came an increased ability to explore space—and create more space jobs. Many of those people inspired by the Moon landing contributed to this book in the hope that their commitment to space exploration is contagious.

Inspiration alone could not get us back to the Moon and on to Mars, though. That takes money, and after Apollo, America decided on channeling more government funding to Earthbound efforts and less to reaching for the stars. Instead of racing into space, we trudged toward it.

After Apollo, the space program was scaled back, to *Skylab*, *Apollo-Soyuz*, and then the space shuttle. Fortunately, we never completely deserted space. The International Space Station (ISS) is flying and being maintained, and experiments in materials processing and life sciences are ongoing. The shuttle program ended, although the Russians kept transporting astronauts and cosmonauts to the ISS, private space transportation companies based in the United States came into the fold to replace the shuttle, and China sent its Yutu-2 rover on an exploratory mission on the far side of the Moon.

We probably could have reached Mars by now if the United States hadn't lost interest decades ago in space exploration and terminated programs designed to support long missions. But this book is not about regrets; it's about the reality that we are now truly on our way to Mars and need a lot of talent and interest to succeed. We need a whole new generation of people inspired to imagine possibilities and innovate as a result.

A LITTLE HISTORY

Leaving other countries out of the discussion for the moment, we could say that none of the delays or shortfalls were caused by the National Aeronautics and Space Administration (NASA). There has never been a US policy to open the space frontier—and NASA can only do what it's mandated to do. Project Apollo was simply "to land a man on the Moon and return him safely to the Earth," and to get there before the Soviets and by the end of the 1960s, no more. After that, the government faced the problems of the Vietnam War and the Watergate scandal, putting the space program on the back burner.

Actually, the US space program started to decline two years before *Apollo 11* landed on the Lunar surface. During the famed race for the Moon, many powerful people in Washington felt dread that, should the Soviets reach the Moon first, they would claim the entire satellite for their own and put up a military base that could threaten Earth.

Members of President Lyndon B. Johnson's administration proposed a treaty banning the ownership and militarization of not only the Moon, but also of any celestial body, by any country. They brought the treaty to the United Nations.

The *Treaty on Principles Governing the Activities of States in the Exploration and Use of Outer Space, Including the Moon and Other Celestial Bodies* became known as the *Outer Space Treaty*, and it stated that the Moon and other planets and asteroids can be used for peaceful purposes only, and that no country can lay ownership to any of these heavenly bodies.

In 1967, the treaty was signed by the United States and the (then) Soviet Union. As of June 2019, 109 countries are parties to the treaty. When it first went into force in October 1967, worldwide fears about a single nation dominating the Moon or other celestial bodies were alleviated, and people breathed a sigh of relief. (It's likely no coincidence that the original *Star Trek* television series and its peace-oriented United Federation of Planets debuted in 1966 as the treaty was taking shape.)

However, the Johnson administration decided to use the bulk of the federal budget for two other purposes, while sacrificing the space program: Johnson's Great Society, and the war in Vietnam.

During the Nixon administration, *Apollo 11* finally reached the Moon. From a public relations perspective, Nixon had no choice. Kennedy remained a beloved president who had made an inspiring promise to his fellow Americans. A country dispirited by an unpopular war and living with the discomfiting fog of Soviet power needed the victory of a Moon landing to feel whole, unified, and positive.

After six more voyages came *Skylab* in 1973, built from the upper stage of a salvaged Saturn V rocket originally planned for *Apollo 20*, which was canceled in January 1970. *Apollos 18* and *19* were canceled in September 1970 due to budget cuts.

When Apollo did land on the Moon, what would follow drove the imagination. Many proposals surfaced, and NASA did show an interest. Multiple versions of the Apollo Applications program hit NASA's radar, and after the first Moon landing, a document called *Report of the Space Task Group—1969* incited a lot of conversation.

The *Report* was for advanced space missions, including a mission to Mars, but Nixon wasn't interested.

There was also the Apollo Applications Program (AAP), using Apollo hardware (the Saturn rockets and an advanced version of the Lunar Module, along with lunar rovers) to launch a space habitat to the Moon, and eventually house six astronauts for six months at a time. Again, Nixon wasn't interested.

Nixon did not want to abolish the space program, but he did want to put it on equal footing with other programs in the federal budget. That translated into a conservative approach to funding. The proposed Apollo Applications Program eventually came out with *Skylab* and *Apollo-Soyuz*.

The more ambitious Space Task Group proposal eventually did turn out the shuttle and later, the ISS, but not in the way they were originally proposed. Nixon accepted the space shuttle program, which became his own project after Kennedy's Moon program. It was President Ronald Reagan, in 1984, who ordered the space station to be built, and it took about twenty years to get it flying with a permanent crew.

The best of the proposed post-Apollo programs, with a fully reusable, bigger than the shuttle rocket, would have been two-staged, with the lower booster stage returning to Earth under its own power while the upper stage would proceed into Low Earth Orbit (LEO). The next step would have been a space station. This station would have been a bigger version of the ISS, not only serving as a way station, but also perform experiments in materials processing, life sciences, and research in astronomy—and eventually housing up to fifty people. As a way station, astronauts would be transferred to another ship, bound for the Moon, where they then would have proceeded to a lunar orbiting space station, also proposed. From the lunar space station, another ship would have transferred them to the surface, where they would proceed to their habitats. All this would have been a multiple-part project, all by NASA: building a shuttle, two space stations, Moon-bound ships, and a Moon base simultaneously.

There was also the Apollo Applications Program, or Apollo X (that's "ex," not 10). Quite simply, the Saturn V would have been used to ship habitats to the lunar surface, and a crewed Apollo spacecraft would have sent astronauts directly to the Moon, with an advance lunar module. A lunar rover would have settled in the habitats for six months at a time. Eventually six astronauts would have ended up being there. There also would have been other uses such as a manned Venus flyby.

Regardless of which program would have been chosen, all this would have been done during the 1970s, and by the 1980s, we might have reached Mars.

Or would there have been other reasons for delay? What if Nixon had been more ambitious for space and backed the proposal by the Space Task Group back in 1969, and if not that, the Apollo Applications Program? Would we be a lot farther in space then we are now?

Perhaps, but after Alastair talked with an expert at the International Space Development Conference 2012, and soon after that, to Rick Tumlinson of the Space Frontier Foundation, he concluded there might have been myriad reasons—unrelated to a political agenda—that impeded progress toward a Mars expedition.

If the Space Task Group of 1969, or even the Apollo Applications Program had proceeded as planned, there is a chance that, regardless of what might have been chosen, the program would have failed spectacularly, especially the Space Task Group proposal. There would be two reasons for this. First, the public was losing interest by the final flight of *Apollo 17*. Second, the cost of these projects would have been monstrous, and Congress, the General Accounting Office (GAO), and the public would not have supported this gargantuan, futuristic effort.

Let's start with the proposed shuttle. Originally, the shuttle would have been twice as large as the one that materialized. NASA asked the GAO for $10 billion. The GAO responded by giving them $5 billion, hence the shuttle we had.

The proposed space/way station, far bigger than the ISS, would have supported fifty people and cost far more money. Remember how the ISS was to cost $8 billion, and ended up costing $100 billion, and rising? Different parts were made by different companies, but NASA had to change the design to accommodate more countries joining in on the project while the cost rose—and the United States

Congress kept cutting back the budget for it, delaying completion. With the proposed way station, the problems manufacturing, launching, and assembling it would have been equal to that the of ISS, possibly higher (and don't forget lunar space station and the two Moon-bound ships, or more).

As for the Moon base, if we ever made it that far, money would have been pouring down a bottomless pit, with costs increasing for Moon-bound ships and base components, all made by different companies, and everybody wanting a big piece of the profit pie. If completed, how many astronauts would have inhabited it, and what kind of work would they have performed? If it was to hold, say six to twelve people, would the government have supported it indefinitely? If so, money from the federal budget would have keep going into this venture, and the costs would have increased year by year. Congress would have started to cut back on funding, and there would have been some serious debates about the base versus other needed federal programs.

Congress would not have put up with it. American taxpayers would not have put up with it.

Most likely, no matter what project they chose, it would have been canceled long before it was to have been completed, either by Congress or by some president after Nixon. Even if the appropriations arguments had resulted in funding for a small lunar base, the funding would likely have dried up.

And don't forget launch costs, especially that of the Saturn V. The Saturn V, though the greatest heavy-lift launch vehicle at that time, was very expensive. Parts of the rocket were even handmade. The cost of a Saturn V, including launch, was about $1.5 billion in 2020 dollars. How much money would the government have had to spend should any ambitious project after Apollo been undertaken?

Costs matter, and neither the government, nor the American people, would have tolerated it for long. Cutbacks would have been made, and private industry would not have stepped in during the

1970s. Reason: Why invest in anything in which your company cannot profit? So, we would have been where we are now. In other words, it would not have made any difference.

A PLAN FOR THE PRESENT

Practically speaking, LEO is now being handed over to the private sector. The ISS is being supported by its various countries, and government-funded ships are still being sent there, but private space transportation systems will lead to privately funded space stations and habitats for tourism and other functions. LEO will soon be crowded with different business ventures.

What is to follow is the Moon and possibly its LaGrange points, beginning with L1 and L2, and then near-Earth asteroids. It is at this point where we will stay for a while, building up an infrastructure. Humanity may decide to venture to Mars before that infrastructure is built, but it would be foolish to remain there for an extended visit until it exists.

The movement to space needs to be restructured. First, many of its functions should be handed over to other government administrations and private industries. That is, the operations of weather satellites should be handed over to the National Oceanic and Atmospheric Administration, and all transportation to LEO should go into the hands of the private sector.

The function of NASA and the government should be the following:

- Longterm research, especially in endeavors private enterprise cannot afford to do alone
- Nonprofit research
- Deep-space exploration

- Situations where NASA leases space, but does not
  yield control

Opening the space frontier—and cultivating the myriad space jobs with good wages and worker protections—involves the partnership, not the dominance, of these three entities:

- Government
- Private enterprise
- Professional organizations, including organized labor

The government will serve as a catalyst, a cheerleader, a lawmaker, an enforcer of laws, and a keeper of the peace.

Many people, including both scientists and politicians, debate the vision in space itself. Some argue robots can do a better job of planetary exploration. Others believe the Moon is a dead end; they think we will never venture to Mars and other worlds if we settle on the Moon, and we've already been there with Apollo.

These people have missed the point! If the Moon is to be explored, settled, and its resources utilized—and that will happen—it is inevitable that we go to Mars, the asteroids, and the worlds beyond that. We are explorers by nature, and once we are on the Moon, we will want to expand further out in space. For anyone who appreciates the school motto *schola munda est* ("the world is our classroom"), every time human beings and their robots venture beyond Earth, our classroom just gets bigger and bigger.

INTO THE FUTURE

We will build our first extraterrestrial civilization by creating settlements and exploiting resources of the Moon, and we will extract resources from near-Earth asteroids. As the lunar settlements grow, more and more people will emigrate. Many will be people with everyday skills like driving, road construction, and cooking—but what will set them apart from other Earthlings is their desire to do those jobs in an extraterrestrial environment.

There have been discussions about going directly to Mars, bypassing the Moon. The arguments for and against are the subject of other books; this book only touches on them. Here we take you on a job hunt, if you will, by taking the slow-and-steady approach. Through stories about industry, university, and government projects, we look at space careers on Earth, in Low Earth Orbit, and on nearby celestial bodies like the Moon. We explore ideas for infrastructure in reaching the Moon for purposes of establishing settlements and then launching to Mars.

The first step is to have a space station in Earth orbit with the sole purpose of transferring people and equipment from Earth orbit to the Moon. The existing ISS is not sufficient for the job, because it is multi-purpose, too expensive, and the process of transferring people and supplies would interfere with the present operations. Later in the book, we'll look at the components of a separate way station and lunar transportation system.

In building a base on the Moon, the process should be simple, inexpensive, and should start with a crew of six to twelve people—taxing and lonely, but probably well-paying jobs. From there, we can expand into a more complex system involving more people. The base would start off as experimental, building habitats for shelter and procuring food sources by way of lunar agriculture. In order for the base to function, it would start off as a laboratory, examining and experimenting with lunar resources: the soil, rocks, ice, and minerals.

Here would begin the concept of In Situ Resource Utilization (ISRU), the process of using lunar resources to expand the habitat by producing building materials on the lunar surface, and developing new products to be sold for a profit. The Moon's one-sixth gravity would make the finished product even purer, because the elements would have a higher chance of mixing together.

It is from here, using ISRU, that the lunar economy will develop. It's at this point where the private sector, both commercial and academic (universities), will want to get more involved. Once the Moon's resources are available for utilization, profit is in sight, giving commercial entities more incentive to invest.

From here, the lunar settlement will expand with bigger habitats, labs, and manufacturing facilities, paid for by the private sector, but also subsidized by various world governments, at least for a while. When a big enough profit is made, these settlements will have the ability to expand on their own. The base evolves into a settlement, that evolves into a city, completely self-supporting, both financially and technically.

In addition to building up a society on the Moon, near-Earth asteroids would also be mined, developing benefits for both Earth and for space habitats, providing what the Moon needs. In fact, companies today are proposing mining asteroids robotically; their results would complement what we can do on the Moon.

Mining asteroids, as well as mining and settling the Moon, could be done simultaneously. Extracting and processing resources from celestial bodies may be the first industry to profit, initially helping to support lunar settlements.

Yes, we are on our way to Mars in this book, but the myriad activities to get us there offer exciting and lucrative opportunities for adventuresome people who are technical and nontechnical, whether armed with PhDs or armed with high school diplomas.

PART I

THE IMPERATIVE

ONE

PROMISE WITHIN REACH

Another ordinary day begins. Mac grabs his gear and heads for the control room. Everyone shows up for the shift who should show up. Nobody sick, nobody injured—because "everyone" except Mac is a robot.

We could be talking about the start of a workday on an offshore drilling rig in the Gulf of Mexico or another day extracting water and precious metals from an asteroid. Regardless of where he is, Mac is making good money doing a somewhat-dangerous job in an alien environment. He and the rest of his team of five humans are away from their families for a month at a time, living in rather comfortable facilities tethered to the asteroid below. He loves the paradox of his life: a daily grind that's a great adventure. Two of the crew are here to install new systems and the other two take shifts with Mac coordinating the efforts of the robotic workforce that drills, excavates, collects, analyzes, and processes resources. Their proximity averts any problems of signal latency in communicating with the robots.

Near-space industry will eventually offer this paradox to women and men with specific skills and guts as well as those with advanced degrees. And the work they do will provide resources like water and materials for off-Earth settlements as well as bring valuable metals such as nickel, iridium, palladium, platinum, gold, magnesium, osmium, ruthenium, and rhodium home to Earth.

The last three elements are not commonly known but, like the others, you see them every day. Osmium was discovered in 1803 and is used to make very hard alloys such as those found in ballpoint pens and electrical contacts. Ruthenium was an 1844 discovery now used in solar cells, turning light energy into electrical energy. Rhodium was another element discovered in 1803 and you're likely to see it used in furnace coils and electrodes for aircraft spark plugs.

A *Wall Street Journal* article on the 2019 trade war between the United States and China noted:

> As trade talks between the U.S. and China have focused on manufacturing, American negotiators seem to be ignoring China's growing domination of raw materials that are crucial to both countries' security and standard of living...
>
> China now controls the supply of all sixteen strategically critical rare-earth metals.[1]

The point is clear. The trade war focused on pork, soybeans, and steel is masking another important trade war—one that is potentially a key to the future of Western economies. It might be won, or cooperatively negotiated, by exploiting the riches of near-Earth bodies such as asteroids and the Moon.

From luxury jewelry to catalytic converters in cars, the minerals that can be mined on asteroids and the Moon are already part of our economy on Earth. We don't just like them; we need them to provide the building blocks of daily life. We need them for clean water and affordable transportation all over the world.

Yet the kind of mining described at the opening of this chapter is a futuristic vision, despite the rapid pace of technological progress in the space industries. Before humans reap the rewards of asteroid or lunar mining, companies with a broader view of space activities must take shape and thrive. It is that breed of space companies—well-established aerospace companies as well as start-ups—that have earned the spotlight in this book. Their work is the foundation for space activities in 2035 or 2040, when we will likely be mining celestial bodies and developing settlements on the Moon into permanent homes.

At this point, any curious person would wonder how it's even physically possible to mine asteroids, which are zero-gravity (zero-g) bodies that neither robots nor humans could stand and walk on as we do on the Moon. (Specific methods receive attention in Chapter 6.) There is another challenge, though: Asteroids are far more rogue in their movement than the Moon, so a first step is to stabilize the position of the target. One possible solution is using LaGrange Libration Points, which refer to positions in our sky having specific properties related to the Moon's oscillation patterns.

There are five libration points of the Moon's orbit around Earth (see Figure 6.1 in Chapter 6), where gravitational and centrifugal forces balance in the rotating system. This is a balance like that achieved by Earth and the Moon or Earth and the Sun. Objects placed at these locations tend to remain there with minimal expenditures of energy.

The orbits known as L4 and L5 are ideal locations to place asteroids to be mined. Those known as L1 and L2 would be perfect locations for way stations for Moon-bound ships or mass catchers for lunar resources. As Neil deGrasse Tyson noted in his book *Space Chronicles*, "Yes, the Moon is a destination. Mars is a destination. But the Lagrangian points are destinations too."[2] An asteroid could be captured and essentially put in satellite status. NASA's vision for accomplishing this has involved unmanned robotic missions (something also explored in greater detail in Chapter 6).

FINANCIAL MOTIVATION, HUMANITARIAN RESULTS

The Apollo Moon landing was perhaps the greatest achievement of mankind. The problem is that it was conceived of and managed by a government that simply wanted to demonstrate it could produce a more dramatic fireworks display than any other nation on Earth. It reflected an overfed space bureaucracy that yielded spinoff technological benefits but had more hope and ignorance in its program than practical goals.

If scientists had known then what we know now, the grand Apollo adventure would have been a direct route to profit—not a bad word if it improves the quality of life for billions of people. But when Apollo was underway, we didn't know that near-space bodies offer unimaginable wealth in the form of minerals. The Moon is also a payload of valuable minerals currently worth trillions of dollars.

Commercial space exploration has the potential to expand the economy as it provides the resources desperately needed by many countries to pull massive numbers of people out of poverty. This is not a utopian fantasy, but rather a realistic vision wherein some people make a lot of money and many people with very little of anything finally have a decent amount of something. Like water, warmth, and the ability to get from here to there affordably.

This doesn't necessarily mean that we will avert resource-based wars, such as those we have experienced over fossil fuels and precious minerals. But it could mean that wars over resources would be less likely.

If you're in the playground fighting over candy bars and a truck-load of peanut butter cups falls into your playground, the battle-ground changes.

Abundant resources is only one of many compelling reasons to plant our feet on asteroids and the Moon, and to build space stations where ordinary people—as opposed to astronauts—can live and work. The four overarching reasons we have targeted are *homesteading*, *research*, *production*, and *profit*. In this chapter, we introduce just a few examples of how activities in these four areas are taking shape.

HOMESTEADING

In the movie *Elysium*, a space station features advanced technology in a "perfect" home for Earthlings of wealth and privilege. The predictable conflict is that people on Earth suffer the effects of a crumbling civilization—and will do anything to get to the space station.

Here's the good news: None of the so-called space barons like Elon Musk, Jeff Bezos, or Richard Branson shows any signs of being the evil overlord that Jodie Foster portrays in *Elysium*. In other words, they aren't trying to design and sustain an escape habitat for rich people. It's true that rich people will be first space tourists—a phenomenon explored in Chapter 8—but their sense of adventure and ability to generate profit for space entrepreneurs is potentially a good thing for all of us.

Here's the bad news: The part of the story about crumbling civilization is where we're headed. We live in a finite system on Earth, with an ever-expanding population, and we are using up finite resources to maintain a high standard of living in the so-called developed nations. Humanity tends toward continual growth of population, land occupancy, production, waste, consumption of energy, and exploitation of resources. Somehow, this has to end, or society will collapse. Even if we end the resource problem by extracting what we need from celestial bodies, we still face the challenge of a growing population. Homesteading on manmade LEO bodies like *Elysium*, the Moon, and then later Mars is one way to alleviate congestion on Earth.

One of the first steps toward near-space habitats may be the kind of environment suggested at the beginning of this chapter—temporary quarters akin to an offshore drilling rig. It's not exactly homesteading, but it is a giant step toward getting used to a frontier environment on a daily basis. After we do that, bringing onboard the equivalent of the military MRE (meals, ready to eat) to sustain life, we will be faced with basic human needs of homesteading: food, clothing, and shelter that not only sustain us, but give us the pleasures that make life worth living.

Of course, we currently don't have all the answers about what the first homesteaders on the Moon or Mars will eat or wear, or the environments where they will sleep. We do, however, have a foundation to project what they will be like. We know the research, inventions, and ideas suggesting options in the early stages of homesteading.

Creative minds from fields ostensibly unrelated to space have been among the most prominent in helping space agencies and companies solve the challenges of feeding, clothing, and housing human beings in off-Earth environments. In this chapter and others, we also look at people with essential, daily life skills that play a key role in getting us into space and enabling us to live there—people like forklift operators, tailors, and physical therapists.

FOOD

You can expect craft beer in your Moon home, but you will only dream about a real hamburger. (More about the beer in the Research section later in this chapter.) Growing plants in space, creating water, and recycling urine into drinkable liquid are logical and familiar processes. No one has come up with a good reason to transport cattle in a spaceship, though.

In 1964, Swiss metabolism expert Max Kleiber told attendees at the NASA Conference on Nutrition in Space and Related Waste Problems what kind of animal would be preferable to a cow for astronauts wanting meat for dinner:

> To produce 7.4 megacalories of human food, which is abundant for two men but may suffice for three, a steer of 500-kg [1102 lbs.] body weight has to be hauled into space. The same amount of food is furnished by 296 rats which weigh 74 kg [163 lbs.] or by 1700 mice with a weight of only 42 kg [93 lbs.].
>
> When weight is important, the astronauts should eat mouse stew instead of beef steaks. [3]

NASA chose to fund efforts to recycle human feces into food rather than develop recipes for mouse stew. In 2018, Pennsylvania State University researchers announced they had developed a recycling system that uses feces to help grow edible bacteria. Their bioreactor breaks down human solid waste into salts and methane gas. The bacteria feast on the gas resulting in a protein-rich paste.

As part of that same 1964 conference where Kleiber presented his mouse thesis, experts discussed how to cultivate plants as long-term food sources for space travelers. They talked a lot about things like algae, endive, and Chinese cabbage. Two and a half years later, a professor of horticulture at the University of Wisconsin named Theodore Tibbitts helped make the first of several breakthroughs for NASA—with beans.

In the early 1960s, Tibbitts kept close watch on the schedule for NASA launches. On the day of a Mercury or Gemini launch, he would grab his family, gather them around their small zenith black and white TV and watch astronauts blast into space. Although his specialty was iceberg lettuce and much of his time was spent in the peat fields of Wisconsin helping farmers increase yields, he held an unusual interest in space, particularly for a plant scientist that worked primarily with tobacco, lettuce, and beans.

In 1965, NASA announced that as part of their plans to go to the Moon, they were going to investigate what would happen if they tried to grow plants in space and film their development. They wanted to move from the speculative discussions of the 1964 conference into a Low Earth Orbit lab.

The NASA operation, which involved the bioscience satellite (Biosat), was looking for horticulturists to work with North American Aviation in Los Angeles for the project. Cultivating plants and filming their growth process were not simple tasks, since soil floats and water forms jiggling droplets when exposed to weightlessness. Serendipitously, Ted Tibbitts was due for his first sabbatical. He applied to the Biosat project, NASA accepted him, and in 1965 Tibbitts moved for a year to Pacific Palisades to help the space agency develop the first spacecraft to fly beans in space, with a movie camera recording how and if they would grow.

After a year in California helping develop the growth systems for beans, he returned to the University of Wisconsin. And then, on December 14, 1966, NASA launched *Biosat 1* and proved that plants would grow in space. The experiment also uncovered a unique phenomenon that, despite the absence of day and night, the bean leaves would wave and bow to the rhythms of the sunset and sunrise. When viewed by time-lapse photography, they looked as though they were waving delightfully in the zero-g environment like a princess in the Tournament of Roses Parade.

Biosat 1 was followed by *Biosat 2* in September 1967 and Tibbitts found himself becoming a space plant expert, with a growing desire to be involved in spaceflight as much as possible. This passion manifested in unusual ways. On his first flight on a 747 returning to Los Angeles for additional research, the plane experienced heavy turbulence. Despite warnings from flight attendants to "Buckle your seat belts and do not leave your seats," he conspiratorially whispered to his ten-year-old son Scott, "Follow me, they won't see us!" He then unbuckled his seat belt, and the two of them sneaked to the back galley of the plane where they couldn't be seen, waiting for wind shear that would cause the plane to drop rapidly. Timing jumps with the drop of the plane, the two of them would float up in the galley for a second or two of weightlessness "just to see what it felt like to be an astronaut." (Scott, who told us this story from his childhood, also found a career in space, and that is the subject of a story related to Mars.)[4]

Thirty years after his jumps in an airplane, Ted Tibbitts was working shoulder to shoulder with astronauts.

In 1988, a team led by Theodore W. Tibbitts at the University of Wisconsin Department of Horticulture published a paper entitled "Cultural Systems for Growing Potatoes in Space." If you've read or seen *The Martian*, you might think of Tibbitts as the one who sustained Matt Damon's life on the Red Planet. Andy Weir, author of *The Martian*, took the sense of what Tibbitts said right up front in his paper and dramatized it accurately as a way of explaining how anyone attempting to homestead on a celestial body would have to rethink how food is cultivated, and what crops could even be considered for cultivation:

> It will be necessary to recycle all inedible plant parts and all human wastes so that the entire complement of elemental compounds can be reused. Potatoes have been proposed as one of the desirable crops because they are 1) extremely productive, yielding more than 100 metric tons per hectare from field plantings, 2) the edible tubers are high in digestible starch (70 percent) and protein (10 percent) on a dry weight basis, 3) up to 80 percent of the total plant production is in tubers and thus edible, 4) the plants are easily propagated either from tubers or from tissue culture plantlets, 5) the tubers can be utilized with a minimum of processing, and 6) potatoes can be prepared in a variety of different forms for the human diet.[5]

You can therefore add *horticulturist* to your list of careers that will be a vital part of the workforce needed to explore space.

But then, the space potato expert is only part of a horticultural/agricultural team that includes people who can operate a skip loader and water extraction equipment. Training to use heavy equipment like a loader would be given through a laborer's training school, whereas people in the water extraction operation would probably be a combination of engineers, technicians, and laborers.

With that in mind, consider the layers of jobs added just to laborer training when the workers operating and repairing heavy machines are doing so on celestial bodies. People familiar with the characteristics of lunar soil and asteroid ice would need to train the staff at schools who are teaching the apprentice skip loaders. Mechanics would need training and practice in effecting repairs in weak gravity environments. And how much redesigning and retooling would have to be done to even prepare the loader itself to do its job?

The challenges for any of the people on these teams would depend on where they happen to be assigned, of course. If the aim is to create conditions to grow crops, then the objective is to head toward water.

Trace concentrations of water on the Moon's surface are present, but water ice might be in the cold, shadowed craters at the Moon's poles. Collectively, the asteroids we know about offer a lot of water—between 100 billion and 400 billion gallons (400 billion to 1,200 billion liters) of water spread among these space rocks. And then there's Mars. Since surface soil is the driest, it is estimated that at deeper levels, the average Martian soil may contain at least 3 percent water. With the proper equipment, all the water and oxygen needed for survival can be extracted from the Martian soil.

First, the teams would have to excavate clay minerals, ice, or a mixture of the two (permafrost). Extremely cold ice is very hard, and permafrost is difficult to excavate.

Second, they would heat the excavated material to a high temperature to liberate the water. This would be easy if they mined polar ice, but since establishing a settlement at or near the polar ice caps makes no sense (at least now), the effort must be concentrated on the difficult task of extracting water from clay or hydrated salts.

All that is needed for water extraction is heat. A skip loader can pick up the soil, load it onto a truck, and dump the soil onto a conveyor belt to go through a water extraction facility, and then store the water in tanks.

If the soil has a 3 percent water content, an extraction system using one hundred kilowatts could produce seven hundred kilograms of water per day, or up to fourteen thousand kilograms if waste heat from a nuclear generator is used. This is more than enough water for drinking and growing crops. Maybe not *all* crops at first, but at least we know we will have potatoes.

CLOTHING

On the low end of sophistication in spacesuits is the futuristic outfit designed by Under Armour for Virgin Galactic's space tourists. On the high end are spacesuits made for more than a suborbital ride. Author Michael Dobson is the rare civilian who owns one.

When Dobson joined the staff of the National Air and Space Museum, the facility had recently been renamed and would soon be relocated to its current location on the Mall in Washington, D.C. Known as the National Air Museum before 1971, when this branch of the Smithsonian Institution added "space" to its name, astronaut Michael Collins was soon brought in to lead it. As the Command Module Pilot for *Apollo 11*, Collins orbited the Moon as his crewmates Neil Armstrong and Buzz Aldrin walked on the lunar surface.

As a junior member of the curatorial staff, Dobson had access to shipments of artifacts coming into the museum. They included dozens of spacesuits that had never been in space, so they were destined for the research collection rather than display. Dobson explains:

Because an Apollo mission required years of preparation, each astronaut received six spacesuits, enough to take care of accidents, modifications, and wear and tear. One of the six would go on the mission; the rest ended up in a big pile in a warehouse somewhere in Florida. And one day, someone at NASA decided it was time to clean up the warehouse. If you're looking to get rid of a few hundred used spacesuits, there aren't a lot of places you can send them. There are rules about this sort of thing. So, they decided to "donate" the suits to the National Air and Space Museum, where I happened to be working.[6]

Dobson mentioned to a senior associate that an *Apollo 7* spacesuit would be welcome in his home and got permission to take it. "Don't sell it" was the only caveat. It now hangs in his Maryland office where he writes books about business issues—including space business. His work includes chapters in *Applied Space Systems Engineering* and *Applied Project Management for Space Systems*.

In talking with Dobson for this book, he mentioned something provocative about that piece of clothing: Playtex, the company known for its bras and girdles, made the spacesuit. The division of Playtex that concentrated its efforts on spacesuits adopted the name ILC Dover in 1971, and it continued in the space clothing business throughout the Apollo missions, beating defense contractors in their efforts to move into the business of zero-g haute couture. In part of its fifty-part series celebrating the fiftieth anniversary of the *Apollo 11* Moon landing, *Fast Company* magazine penned a succinct description of the clothing challenge that Playtex met:

One of the underrated technical challenges of going to the Moon was designing the spacesuits. The suits had to be inflated and pressurized from the inside—meaning, they had to carry around a tiny version of the atmosphere human beings require to stay alive. The suits were, in essence, sophisticated balloons.[7]

Given that the suits had to flexible as well as tough enough to withstand a temperature range of about −280° to more than 240° when astronauts were in the Sun, the design challenge was daunting. The astronauts needed freedom of movement so they could climb, twist, and collect samples—all the while facing the possibility of being hit by a micrometeorite going 36,000 miles per hour.

In the lead-up to the *Apollo 11* mission, Playtex/ILC Dover's workforce swelled to about a thousand employees, then dropped to a mere twenty-five after the Apollo missions ended. But ILC Dover made a comeback in devising a way to build suits with interchangeable parts for the space shuttle missions. They then went on build the impact bags to protect the first Mars rover during its landing and contribute to space habitats, among other things. The common thread is the company's use of flexible materials to keep people and equipment safe.[8] As of this writing, some of the positions open at the company are genuinely space-age in nature—Team Assembler, Space Suit Assembler—and some of them seem like a throwback to a pre-industrial era—Sewing Professional.

THE ONLY SPACESUIT COMMISSIONED BY A CONSUMER

Dr. Alan Eustace, former senior vice president of Knowledge at Google, holds the world record for the highest-altitude skydive. On October 24, 2014, he jumped from the upper regions of the stratosphere, leaving his gas balloon at 135,889.108 feet (41.419000 kilometers; 25.7365735 miles). Eustace is not an astronaut; he's a computer scientist.

His was the first spacesuit that had ever been designed for and sold to a consumer. Every other suit had been sold to a government agency. Yet the company that built his suit also built those worn by all the Apollo astronauts. Basically, they knew what he needed—and then some.

The characteristics of his suit, and his body's responses to stratospheric conditions with only the protection of that suit, give unique insights into high-altitude human flight. The stratosphere is not considered space, which is generally marked by the Kármán Line, around 62 miles (100 kilometers) above earth's surface. The stratosphere hosts some remarkable differences from the atmosphere below it and those differences affected the design of Eustace's suit.

The spacesuit was made by ILC Dover, the company that made all the Apollo and extravehicular activity (EVA) suits. They are distinguished from other spacesuits that NASA has because they are meant to be used when pressurized. In contrast, suits made by the David Clark Company were used by shuttle astronauts and normally depressurized. If there is an accident causing a loss of pressure, they will pressurize themselves.

ILC Dover suits are not only pressurized, but they also support mobility. As Eustace says, "They have beautiful joints inside and other features that make the suit comfortable and functional. Your hands turn and you can move your elbows and your shoulders. And you can walk like the Apollo astronauts did."[9]

Although there has been R&D related to spacesuits since Apollo, the prototypes did not tend to make it to the manned-test phase. The Eustace suit was one of the first ILC Dover suits in decades that that went from design to fabrication to pressure suit tests, to human pressure suit testing, to actual use in the upper stratosphere. Using it for the record-breaking jump was a milestone from a scientific point of view.

A key feature differentiating it from predecessors was the level of pressurization. It had the highest pressurization of any US suit ever produced. Most of the suits were pressurized at 4.3

PSI, but the Eustace suit was pressurized at 5.4 PSI. It doesn't sound like that much, but the decompression sickness (DCS) risks went down dramatically with the PSI increase. Given that Eustace never suffered effects of decompression sickness on his ascent, the strategy worked.

As often is the case, however, when one aspect of a product is tweaked to space-qualify it, another feature is affected. With the added pressurization, the mobility of the suit changed slightly. It became a little less maneuverable than before, but not much. Overall, matching the suit architecture with the application Eustace and his team had in mind for the stratosphere jump did not impinge on the integrity of the suit. He ended up with a suit equipped with the life support systems he needed and the physical sensation a scuba diver might have.

The magic in the suit architecture was how the mask and suit worked together in pressurizing the suit, with the pressure then stabilized. When he got into the suit, it was deflated. Oxygen then flowed into the suit and he breathed and exhaled normally. Initially during this process, a valve called a dual suit controller was open. The system would take the carbon dioxide he expelled through a mask that provided his oxygen on inhaling, and then route the CO_2 and the moisture to the lower half of the suit and vent it. So anything he breathed went to the lower half of the suit and was released.

When he wanted to pressurize a suit, he would shut the valve. Every time he breathed out, rather than air going out of the suit, it pressurized the suit up from the inside. Every exhale made the suit get bigger. This is not a common suit architecture, but one that matched his needs for the mission.

"Pressurizing the suit from the inside is a strange feeling. It's like blowing up a balloon from the inside and it doesn't take very long. It only takes maybe a couple of minutes for it to fully pressurize and it stops pressurizing once it gets up to 5.4 PSI.

The dual suit controller vents at that point so the suit never gets above 5.4 PSI."[10]

Protection also matched what NASA's Apollo and EVA suits offered. All the different layers in the suit included protection against micro-meteorite strikes and radiation. The different layers were roughly identical to what astronauts wore during Moon missions. Eustace said, "If you put my suit next to Neil Armstrong's suit, there are way more similarities than there are differences."[11]

These suits are very expensive, not necessarily because the materials are expensive, but because very few are ever made. Considering the R&D that goes into them, the company is justified in charging a lot of money for them just to recoup the investment. Eustace notes that when you buy the suit, you also engage suit designers and technicians involved in every aspect of engineering the suit and testing it, so you are buying talent and skill in addition to a product.

One critical system the ILC Dover team needed to monitor was the heating-cooling apparatus. Previously, suits were never heated; they were made to be air conditioned due to atmospheric conditions. Eustace required something opposite; he wanted to be able to heat it because he was going into very cool temperatures.

He expected −120 or more degrees Fahrenheit. He got a surprise. The phenomenon is explained by the Center for Science Education of the National Center for Atmospheric Research/ University Corporation for Atmospheric Research:

"Ozone, an unusual type of oxygen molecule that is relatively abundant in the stratosphere, heats this layer as it absorbs energy from incoming ultraviolet radiation from the Sun. Temperatures rise as one moves upward through the stratosphere. This is exactly the opposite of the behavior in the troposphere in which we live, where temperatures drop with increasing altitude."[12]

In Eustace's words, "We designed something that would allow me to stay warm even in a really horrible environment that we simulated on the ground, but in truth, the stratosphere is actually a pretty nice and balanced place for a human body to live. It was a surprise to us. I actually turned off the heaters at 70,000 feet and never needed them again. And was perfectly comfortable in the stratosphere."[13]

The final point is that the suit was incredibly durable. After a dozen crash landings, Eustace and his team documented that the suit strengthened rather than degraded. The design and execution enabled improvements in sealing connections and reinforcing durability.

Photo courtesy of Alan Eustace, taken by Volker Kern on the morning of the successful completion of the highest-altitude skydive ever attempted. The combined weight of Eustace, the suit, chest pack, and parachutes was around 405 pounds.

Playtex/ILC Dover's partner in technology, though not an actual corporate partner, was DuPont, which developed the science essential to the integrity of twenty of the twenty-one layers of the Apollo suits.[14] The spacesuit related products the company featured in its

"Suited for Space" Smithsonian exhibition included the following—many of which will sound familiar because of the Earthbound uses we have found for them:

- DuPont™ Kevlar® fibers, used to provide strength and flexibility in spacesuits, are used in bullet-resistant body armor to protect law enforcement officers, first responders, and the military.
- DuPont™ Nomex® fibers, used as strong protective layers in spacesuits, are used in garments to protect firefighters, soldiers, and race car drivers.
- DuPont™ Kapton® polyimide film, used in two layers of the Apollo suits because of its durability and thermal stability, is a critical material for high reliability in the electronics industry.
- DuPont™ Krytox® performance lubricants, first used by NASA for the Apollo space flights, in the traction motors on the lunar rover, and as a lubricant for spacesuits, work at both low- and high-temperature extremes to protect everything from computer chip manufacture to industrial and automotive applications to the latest civilian and military aircraft.
- Mylar® polyester film, used in several layers of Apollo spacesuits because of its toughness and flexibility, is used in diverse applications for the electrical, electronics, industrial specialty, imaging and graphics, and packaging markets. Mylar covered with foil is used for balloons with a float time that far exceeds that of latex balloons.

Next-generation versions of these spaceworthy materials have been emerging through the years and they suggest what would be worn on the way to and from space stations and habitats, as well as on any extravehicular activity (EVA).

Clothing worn on the International Space Station (ISS) indicates what will be worn by homesteaders once they get to their destination. Astronauts and cosmonauts on the ISS have had their choice of Russian- or US-made work clothes—coveralls and underwear made of the same types of fabrics they might wear on Earth. ISS maintains an air pressure of one atmosphere, which is the same as Earth, and a comfortable temperature and humidity so normal clothing works except for launch and reentry or talking a walk outside. The big difference is that ISS inhabitants have only an extremely limited ability to wash their clothes. They focus on keeping clean underwear and then put their dirty coveralls into something that allows them to be incinerated upon reentry into Earth's atmosphere.

By the time we have homesteaders on celestial bodies, dirty clothes will not be a problem. NASA is already pursuing development of a washing machine suitable for reduced-gravity environments.[15]

SHELTER

The Space Shuttle introduced a new type of space vehicle. It was reusable and had a wide variety of purposes: launching and retrieving satellites, carrying space laboratories and habitats, aiding in construction of space stations, and transporting personnel. With the payload constraints of the shuttle in mind, as well as those of rockets launched by private companies, creative thinkers considered how to get habitats for humans from Earth to a celestial body.

In 2016, NASA selected six US companies to develop prototypes for deep-space habitats. The competing designs had to be compatible with plans for NASA's proposed Lunar Gateway, a space station serving as a communications hub and short-term habitation module. (Note: We think the Gateway as conceived is a flawed idea and explain why in Part IV.) The current plan is that a bare-bones version of Gateway would be launched as soon as possible so it could be used as an interim station where astronauts could transfer from their spacecraft to a lander, which would then take them to the Moon by 2024.

The different and converging approaches to habitats suggest viable models that may eventually house humans for both short-term missions and long periods of time. Two of the proposed corporate models for solar-powered habitats involve inflatable technology, with two others repurposing existing equipment and one arriving in space with some assembly required. Details of the six habitat designs are covered in Chapter 7 along with alternative concepts for shelters on the Lunar surface and on Mars.

NASA ended the competition on July 19, 2019, concluding that NGIS was the only contractor "with a module design and the production and tooling resources capable of meeting the 2024 deadline."[16] The winning concept was a habitat small enough to launch on a commercial launch vehicle. The construction would occur on a cargo spacecraft called *Cygnus*, which was already in production. (In October 2014, Orbital Sciences *Cygnus CRS Orb-3* blew up on the launch pad in its fourth attempt to reach the ISS. However, the automated spacecraft is now built by the new owner, Northrop Grumman.)

Astronauts on their way to the Moon, researchers, tourists, and probably even a few writers would be housed in facilities like these on a short-term basis. The plan is to extend the duration by adding modules to the basic Gateway that would enable relatively long-term occupation. But even though the length of a stay would not be long in the first phase of development, the common intent with all the designs is that inhabitants be comfortable to whatever extent possible. The models provided to NASA for Gateway consideration foreshadow what we can expect in the design of more permanent facilities for homesteaders.

One required feature of any habitat is shielding it from radiation, which would destroy any living being that is unprotected. There are a couple of ways to do this with locally sourced materials:

- Water surrounding a habitat would protect the inhabitants from radiation. Water traps the high-energy particles, so having about a foot of water circulating around the exterior of a habitat would provide the protection required by people, plants, and animals.
- Lunar soil could also be used to cover the habitats for homesteaders to shield against radiation. Only two meters of lunar soil is required for protection from solar and cosmic radiation, occasional solar flares, and temperature extremes.

This regolith can also be exported to advanced space stations for use as shielding. Lunar rocks can be crushed to a suitable coarse of aggregate size, mixed with cement paste, and form lunar concrete. Reinforcing it with steel or glass fibers increases flexural strength and restricts the growth of micro-cracks. Strong in compression, the lunar concrete, with sealant on the interior surface, can be used to build lunar structures.

Lunar bases made of concrete offer the following advantages:

1. Energy ratio between aluminum alloy and concrete is 90:1, saving massive energy costs.
2. Concrete can be cast into any monolithic configuration.
3. It has great strength.
4. It has first-rate heat resistance.
5. It provides excellent radiation shielding.
6. Concrete provides abrasion resistance, especially against micrometeorites.

Concrete can be a major component in expansion of a lunar base and can be produced on the lunar surface. The only imported ingredient needed is hydrogen, to be used for water. (Any hydrogen present in water vapor is lost to outer space when the water vapor is decomposed by sunlight.)

RESEARCH

The International Space Station (ISS) has been home to thousands of research experiments in biology and biotechnology, Earth and space science, human research, physical science, and technology. All of them have been designed to help us understand life in and around celestial bodies or to give us a different perspective on the problems we face on Earth, such as diseases. Contrary to what some scientists have stated, the ISS can continue to be very useful, perhaps especially to companies in their research and development programs. It could even become an industrial park, with many different companies using it to test new products they may someday market. They might also use the ISS as a small-scale manufacturing plant for these products. So far, about $100 billion has been spent on the ISS; that should be incentive enough to find an ongoing use for it.

A cursory look at the type of research that can only be done in— or is best suited to—a zero-g or weak-g environment such as the ISS or the Moon points to huge opportunities for medicine, agriculture, transportation, and much more. NASA has produced a handbook for commercial and academic researchers interested in working in microgravity or zero-g and outlined the advantages of the environment to them. In the section on "microgravity benefits for material systems," the booklet cites the following:[17]

- Defect free
- Homogenous
- Controlled, symmetric growth
- Avoidance of nucleation or single nucleation
- Higher resolution
- No solute buildup
- No sedimentation
- No convection
- Containerless processing

- Free suspensions
- Perfect spherical shape
- No wetting

The NASA Ames Research Center team authoring the study provided similarly compelling benefits for the life sciences.

SingularityHub published an article about "wild projects" on the ISS that suggests precisely the kind of scientific research that is best conducted in off-Earth environments.[18] Note the fascinating—and in some cases, unearthly—jobs associated with some of the "wild projects":

- **No-gravity environment for Parkinson's research.** Actor Michael J. Fox, who has Parkinson's established a foundation for research on the disease and partnered with the ISS on "research targeting a protein produced by a gene mutation believe to be involved in Parkinson's."[19] Chemists working on drug therapies targeting the suspect protein lack insights into its crystalline structure, but the zero-g environment allows crystals to grow to a larger, more uniform pattern, making them easier to study.
- **Tissue chips in space.** Tissue chips contain human cells grown on an artificial medium. Because the cells imitate the structure and function of organs and tissue, they can lead to a better understanding of the role of zero-g or weak-g environments on human health and diseases. With the objective of improving human health here on Earth, tissue chip projects emulate the complex biology and mechanisms of an organ outside of an animal for drug-screening and other applications.
- **Brewing beer.** Well, sort of. Anheuser-Busch, most known for the Budweiser line of beer, provided barley seeds—important in brewing Bud—for ISS experiments to see how the grain responds to microgravity. Space Tango, a company founded in 2014 by Twyman Clements, supported this experiment. A

profile of Space Tango's projects, aimed at taking raw material and working with it in microgravity to bring new products to Earth, is included in Chapter 6. Space Tango's work exhibits the breadth of commercial opportunities in multi-disciplinary areas available immediately in space-based R&D.

- **Cancer research.** Due to his research role with BioServe Space Technologies, which is part of the University of Colorado, Dr. Luis Zea has been among those designing cancer-related experiments for the space station. Zea is not an MD, as researchers designing medical experiments on Earth might tend to be. He has a PhD in Aerospace Engineering Sciences—Bioastronautics. He explains a key reason why cancer research in a microgravity environment overcomes challenges of doing the same experiments on Earth:

> We can grow 3D tissue in a fashion that better resembles what happens in the human body. Cancer tumors grow in different ways in the human body. It's a 3D structure, sometimes almost spherical. That's what happens in-vivo, in the human body.
>
> When you try to do that in-vitro or in a test tube or a petri dish, it looks more like layers. Just looking at the shape, you could compare it to something that looks more like a tater-tot in 3D dimensionality versus a pancake. Although it's the same material, just the fact that it has a different shape changes a lot of things. So if you don't have the right shape, then you can't really conduct the test as accurately as you would like to.[20]

With microgravity allowing researchers to maintain the integrity of the shape, it becomes more likely that scientists will be able to pinpoint cellular changes that cause cancer. With all of us, at some point, being affected by cancer in ourselves or in a family member, the relevance on Earth of this kind of research is one more powerful justification for lifting off the planet.

- **Experimental chondrule formation.** This discussion introduces a career that most people have never heard of: cosmochemist, a person who explores the mystery of chondrules and other things related to the chemical composition of and changes in the universe. As the demand for cosmochemists increases, hopefully the salary for these PhDs will also; as of 2019 the average salary for postdoctoral jobs was only $73,232, according to information supplied by employers to the website SimplyHired that year. Chondrules are tiny spheres embedded in meteorites that some scientists believe are the building blocks for planets and moons. The experiments in chondrule formation mimic probable conditions that form chondrules, hopefully discovering something about how Earth was created. That knowledge would help in our search for other habitable worlds.

Versions of some of the experiments above have, of course, been attempted on earth. Some research and development could not be, however, because of the potential hazards of attempting these efforts on the planet we currently inhabit. One example is the R&D associated with nuclear thermal rocket systems. The aim is devising transportation technology that dramatically reduces the time to get humans from Earth to Mars so astronauts don't spend nine months to years (or more) relying on an advanced life support systems; requiring a huge quantity of food, water, and fuel; and taking risks with dangers like cosmic radiation bursts.

A nuclear thermal rocket uses a nuclear reactor to heat propellant to a high temperature. The propellant is then expanded by a supersonic nozzle to produce thrust, much in the same manner as a conventional rocket engine. A low molecular weight propellant may be used, such as hydrogen heated to high temperatures.[21]

A nuclear thermal system can be useful for a second-generation upper stage. This concept is also attractive for the human interplanetary phase of Earth-to-Mars transport. We explore this in much greater detail in Chapter 5.

PRODUCTION

Unlike human flesh and bones, certain materials are ideally suited for functioning in space. In addition, the conditions of space can allow for refinements in manufacturing processes, reducing defects and increasing performance, as well as manufacturing with materials that have different characteristics in zero-g or low-g environments. There will come a time when exports from off-Earth manufacturing facilities will be common, so consider an MBA in export management as good preparation for a space-related career.

One of the first products that will likely be exported to Earth is ZBLAN, so-named for its chemical formula ZrF_4-BaF_2-LaF_3-AlF_3-NaF. The modern digital world relies on threads smaller than a human hair that can transmit light pulses of energy at billions of pulses per second over vast distances; in other words, we have a growing dependence on optical fibers. Commonly, they are now made of silica glass, but the signal losses associated with that kind of fiber optical thread necessitate expensive repeaters. In contrast, ZBLAN is a fluoride glass optical fiber potentially capable of ten to one hundred times lower signal loss than silica fiber. Unfortunately, it has unrealized potential when made on Earth:

> When ZBLAN is produced on Earth, convection and other gravity-driven phenomena can cause imperfections because of the nonuniform distribution of the various chemical components within the fiber. These defects that occur during the process of solidification

result in the formation of microcrystals that render the fibers unusable for many commercial applications. [22]

Scientists avoid the adverse effects of gravity by having ZBLAN fiber produced by the ISS National Lab.

Commercial companies such as Fiber Optics Manufacturing in Space, Made in Space, and Physical Optics Corporation jumped into the opportunity to pursue LEO production of ZBLAN and are doing so on the ISS. Considering how dependent Earthlings have become on speedy internet and voice connections, the revenue from their efforts could be impressive despite the hard and soft costs of launching the raw materials into space, which currently come to "thousands of dollars per kilogram," according to Alex MacDonald, a senior economic adviser for NASA.[23]

The Moon's low gravity can be utilized to make a number of finer and better products than those manufactured on Earth. In addition to ZBLAN, many others can be included in the list of products that could be produced with fewer defects and higher efficiency in a low-gravity environment. They include finer glass, purer chemicals, purer medicines, crystals for computers, and alloys for various purposes.

Manufacturing and recycling in space will also involve products and equipment intended to remain off-Earth. Founded in 2010, Made in Space calls itself "the space manufacturing company" and its partnership efforts with NASA have brought that tagline to life. They develop additive manufacturing technology that operates in microgravity. The first effort aboard the ISS involved their 3D printer, which enabled astronauts to manufacture parts on board—a less expensive alternative to shipping them from Earth. Made in Space is now developing a robotic system called Archinaut to assemble hardware autonomously, making it possible to build satellites and instruments in space. After the success of the initial 3D printer, NASA teamed up with Tethers Unlimited, another commercial venture in additive manufacturing, and brought the Refabricator on board the space station. It's an integrated 3D printer and recycler that helps astronauts turn plastics on the space station into 3D printer filament, so they have the material to effect repairs.

PROFIT

There is a way to pay for mining, transportation, homesteading and other near-Earth activities without getting citizens knee-deep in tax debt—which is what would happen if space explorations and development were left to government bodies to pursue. The project can pay for itself. Using the Moon's natural resources, along with the resources of near-Earth asteroids, we can produce marketable products needed both in space and on Earth. The process not only lessens the cost of the Moon/Mars project but may also generate enough profit to strengthen Earth economies.

The rush to profit would have at least two phases. The first phase involves companies willing to take a high risk with the promise of a huge return on investment. In the second phase, competitors to the pioneers will emerge—mining sites on celestial bodies, manufacturing and exporting products to Earth, and so on. They face less financial risk but will also profit less because their efforts will help drive down the price of the precious metals and other items they deliver to Earth. That's more than okay—that's fabulous! Not only are adventurous companies still making money in phase two, but people on Earth benefit from a drastic reduction in the cost of resources like platinum group metals, which seem to be the key to pollution-free cars and clean energy technologies.

During the early days of asteroid mining, to name just one industry, it will be possible for companies to achieve almost incomprehensible wealth despite the costs of capturing an asteroid to low earth orbit (or taking it to one of the LaGrange Points) and building the infrastructure to process the resources in it. Texas senator Ted Cruz; Peter Diamandis, founder of the X Prize competition to encourage technology innovations; and global investment bank Goldman Sachs are the among the diverse voices that have declared, "The first trillionaire will be made in space."[24]

In early twenty-first-century dollars, what are the costs and revenues that potentially result in such astronomical profits? Starting with costs in the billions, we can, as the "experts" cited above predicted, look toward profits in the trillions—if not more.

Prepared by the Keck Institute for Space Studies (KISS) at the California Institute of Technology, a feasibility study in conjunction with NASA concluded that asteroid mining is feasible. The study explained the process of capturing an asteroid and exploiting its resources:

> The Asteroid Capture and Return mission—the central focus of the KISS study—blueprints the technological know-how to moving an asteroid weighing about 1.1 million-pounds (500,000 kilograms) to a high lunar orbit by the year 2025. The mission's cost is expected to be $2.6 billion.[25]

Now for a moment, let's get crazy with profit possibilities suggested by the Statistica Research, a team of hundreds of research experts whose sole purpose is to collect, digest, and share data. Based on what scientists currently know about asteroids—and, yes, they have names—a rock known as Davida is estimated to be worth 26.99 quintillion dollars.[26] (Note: A quintillion is a thousand raised to the power of six; that's a number with eighteen zeroes after it, whereas a billion has only nine zeroes after it, and a trillion has twelve.) The justification for assigning such high value is rich deposits of materials such as nickel, iron, cobalt, water, nitrogen, hydrogen, and ammonia. It is not alone in the assessment of dollar-value in the quintillions, however. Fourteen other asteroids in the belt are estimated to have resources valued at between 2.5 and 7 quintillion dollars.

To conclude the chapter, we ask you to consider profit as yield or return that goes beyond the spreadsheet—beyond the dollars or euros potentially earned by audacious entrepreneurs. The KISS study drew conclusions about profit for humans that is potentially immeasurable. According to the study, some of the results could be:

- Such a venture represents a new synergy between robotic and human missions in which robotic spacecraft retrieve significant quantities of valuable resources for exploitation by astronaut crews to enable human exploration farther out into the solar system.
- Water or other material extracted from a captured volatile-rich near-Earth asteroid could be used to provide affordable spacecraft shielding against galactic cosmic rays. The extracted water could also be used for propellant to transport a shielded habitat.
- This undertaking could jump-start an entire in situ resource utilization industry. The availability of a multi-hundred-ton asteroid in lunar orbit could also stimulate the expansion of international cooperation in space as agencies come together to determine how to sample and process raw material from an asteroid.[27]

Beyond the monetary benefits of asteroid mining ventures such as those proposed by companies, therefore, there are huge opportunities—and incentives—for multinational cooperation in scientific discovery. An optimist might even say this could lead to more harmonious interactions among humans on Earth.

A DECADES-OLD VIEW OF ASTEROID MINING: RELEVANT STEPS, SHORTER TIMELINE

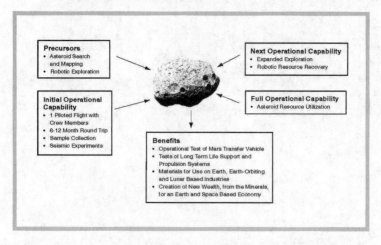

Precursors
- Asteroid Search and Mapping
- Robotic Exploration

Initial Operational Capability
- 1 Piloted Flight with Crew Members
- 6-12 Month Round Trip
- Sample Collection
- Seismic Experiments

Next Operational Capability
- Expanded Exploration
- Robotic Resource Recovery

Full Operational Capability
- Asteroid Resource Utilization

Benefits
- Operational Test of Mars Transfer Vehicle
- Tests of Long Term Life Support and Propulsion Systems
- Materials for Use on Earth, Earth-Orbiting and Lunar Based Industries
- Creation of New Wealth, from the Minerals, for an Earth and Space Based Economy

Illustration by Charlie Shaw. Based on data in "America at the Threshold: Report of the Synthesis Group on America's Space Exploration Initiative," US Government Printing Office, Washington, D.C., 1991, p. A-37.

When your skeptical friends ask the question, "Why should we spend the money and technology to explore space when we have so much poverty and disease on Earth?" you tell them the answer is in the question. We are going into space *because* there is poverty and disease on Earth. We are going into space because the imbalance of resources on Earth may end life as we know it if we don't. We are going into space to meet acute needs in medicine—cures for incurable illnesses and new ways of fixing injured bodies.

Space exploration is really about the quality of life on Earth, at least for the foreseeable future.

TWO

WILL HISTORY
REPEAT ITSELF?

So far, we have optimistically ventured into examples of what the pursuit of profit, as well as homesteading, research, and production might look like if we all (sort of) get along. If we don't, there will still be plenty of jobs in space-related activities, but proliferation of them in sectors of the celestial economy linked to fierce competition, manipulation of market forces, and military and law enforcement strength.

Each of the four areas of activity we explored in Chapter 1 has particular challenges that have been faced before—sometimes badly, sometimes well. With that in mind, let's do some time-traveling, paralleling historical events with evolving issues.

COMPANIES STRONGER THAN NATIONS

Skepticism about humankind's ability to privatize space operations and concurrently create equal opportunities for employment is justified. Skepticism about Earth nations' ability to cooperate on space matters and abandon chances for military and economic superiority is also, of course, justified. Space is not necessarily going to bring out the best in human beings as *Star Trek*'s creator, Gene Roddenberry, would have wished.

To begin, the rise of the British Empire suggests how privatization of certain activities opens worlds of opportunity for entrepreneurs and adventurers that did not exist when government was in control of them. It also suggests how commercial entities can have a moral compass magnetically controlled by profit.

At its peak, the British Empire took up a quarter of Earth's land mass and ruled roughly 20 percent of the world's population.[1] Commerce within that empire involved people migrating from one place to another, yet they were united with an infrastructure and language. Although the empire is now gone, the Commonwealth and the widespread use of the English language remain.

But it wasn't the Crown or Parliament that funded and controlled interests of all kinds around the globe, with the result being an empire composed of fifty-seven colonies, dominions, territories, or protectorates.[2] The empire was not built by the British government, but by entrepreneurs. From helping to spawn super-sized democracies like Canada, the United States, and India, to cultivating and sustaining opium addiction in China, British companies tried to take over the globe.

As the world in the 1600s and beyond was being explored, and the sea lanes opened up to shipping, companies emerged and seized the dominant role in British expansion. These companies were out to make a profit by sailing to different parts of the world, obtaining goods and raw materials, and then shipping them back to England to sell to the English people—or people elsewhere within their reach.

The Hudson's Bay Company began its commercial presence in North American by setting out to Canada and the Great Lakes region to trap animals with valuable furs in 1670. They later set up a fishing enterprise off the Grand Banks, in Newfoundland. HBC's supply ship *Nascopie* began a practice that will no doubt be followed by companies with a commercial footprint on celestial bodies: taking tourists on supply runs.

The British East India Company made small settlements in India to obtain tea, spices, cotton, silk, indigo, saltpeter, and opium. The company later went to China to sell the opium, which was illegal in China, resulting in the deterioration of an entire country and eventually, the Opium Wars when China tried to fight back. Among other things, this led to the British seizing Hong Kong. From this port, the British were able to obtain Chinese porcelain and silk.

The origin of the company was private enterprise in full bloom. In 1600, Elizabeth I granted the company a royal charter, which gave them a monopoly on trade to the East Indies. With that potent authorization in hand, the founders lured wealthy merchants who funded the venture with nearly £70,000 of their own money. According to *National Geographic*,

> Think Google or Apple are powerful? Then you've never heard of the East India Company, a profit-making enterprise so mighty, it once ruled nearly all of the Indian subcontinent. Between 1600 and 1874, it built the most powerful corporation the world had ever known, complete with its own army, its own territory, and a near-total hold on trade of a product now seen as quintessentially British: Tea.[3]

The company having its own army may be the scariest part of this entrepreneurial success story. It used force to squelch rival traders, and ultimately seized control of a state in Bengal, where a company "executive" began collecting taxes. The company later annexed other territories and built an army of 260,000—twice the size of Britain's standing army. It was also responsible for almost half of the country's trade.[4] That could never happen now—right? We have antitrust laws that theoretically would prevent such domination. This theoretical protection gets a closer look Chapter 3.

As for the issue of an army, any company extracting precious minerals, conducting tourism, transporting goods and people, or handling construction of habitats would require a security force. Even if robots were doing all the work, they would still require such a unit (probably made up of intelligent security robots). Depending on the nature of their business, some companies might invest heavily in a security force, creating a kind of corporate army in space. For a moment, we might want to think of Britain as Earth and the subcontinent as the Moon or Mars: With the base of operations so far away, it seems reasonable it would be easier to amass a large "security" force, as the East India Company did.

The Virginia Company, eyeing North America, first settled in Roanoke, but that colony was lost. They then established a post in Virginia with tobacco plantations, exporting tobacco back to Britain for consumption. Known as a joint stock company, the Virginia Company was funded by investors who expected settlers in the New World to send back all kinds of highly valued products—gold, wine, rare fruits, and precious stones. They didn't get all they'd hoped for, partially because they just weren't sure what they would find. A similar company going to an asteroid would have a much better chance of getting what they came for on behalf of their investors.

The Levant Company, formed in 1592, had no colonial aspirations, but their charter involved maintaining trade and alliances with the Ottoman Empire in Turkey. They also traded with existing Middle Eastern ports, in commodities such as raw silk, cotton, wool, yarn, nutmeg, pepper, indigo, galls, leathers, soda ash for making glass and soap, and gums and medicinal drugs. This increased international relations, and may have played a role in the British later establishing protectorates in the Middle East and North Africa. Among the famous wealthy families to emerge from the company, which was dissolved in 1825, were the Astors. We can only guess what new breed of super-rich could come out of successful exploitation of the resources on celestial bodies.

As the posts and "factories" (where a "factor" or trade agent bought and warehoused material for trading ships to load on their arrival) were established at seaports and trading centers to obtain all the commodities and materials related these companies and others, European settlements grew in and around the ports. People who worked at these posts and ports needed a place to live, whatever their occupation. This attracted the attention of Britain's populace. These settlements provided opportunities for work and a new life.

Many groups in England were persecuted for religious reasons at various times, or belonged to groups the Crown wanted to get rid of: Puritans, Quakers, Catholics, the very poor, and criminals (who were later sent to Australia). Many people came to the colonies hoping for religious, economic, and political freedom.

With posts and ports established, they needed more people as a support system in order to make more money, and to provide land to those who were willing to come and deal with the harsh wilderness and the natives. Eventually, this built up into towns and cities as more ports were established to transport goods.

Stations were built in the colonies and on islands in the Caribbean and the Pacific to service and resupply ships.

The British government came, to protect ships and settlers from competing Dutch, Spanish, and French empires as well as local threats. They also sought to assert more authority over settlers, to ensure that money from land grants and claims flowed to the Crown and to landlords, and eventually to establish a rule of law.

Trade grew between far-flung colonies and existing empires, opening up access to goods that previously had been available only to a handful of elites. Seaports and stations grew up on islands to service and resupply ships. Migrants settled to establish new lives, new businesses, or to escape persecution. Criminals were exiled, land was granted to powerful courtiers who then sold rights or granted access to entrepreneurs. Missionaries went out to Christianize natives in distant and remote places.

The British, French, Dutch, Spanish and Portuguese empires all started out this way. Merchants ventured out, followed by colonists seeking a better life, or at least a more adventurous one. This is exactly the way it's going to be as we expand out into space.

New careers will be vital in expansion, such as those that took shape during the global outreach of British companies while they built the empire. But what kind of careers will emerge in the course of dealing with human activity in space if all we do is replicate the complications, aggravations, and injustices we created on Earth?

TRADE WARS REDUX

China had economic dominance in certain products and processes from roughly 1100 to 1800. New types of jobs and economic opportunities came out of their innovations in manufacturing, textiles, agriculture, and much more. But there wasn't really direct Chinese shipping to Europe. Most Chinese goods that reached Europe came overland through the Ottoman Empire—which was why the Portuguese got such enormous profits when they first opened up Eastern markets starting with India. China wasn't especially interested in trade with Europe, which didn't have much to offer: The British cultivated opium in India in order to have something the Chinese would buy. (Chinese consumers preferred opium from India to Chinese-made opium.) Access to products resulting from long-distance trade becoming technologically feasible helped spur economic development in Britain and Europe.

The brute force practices of Western imperialism hacked away at China's ability to remain dominant, or even competitive, in world markets. The 1900s brought military actions, confiscation and exploitation of resources, and diversion of trade that brought China to its knees economically.

Obviously, China is back as an economic force. This time, however, it's not a matter of outpacing Britain in the production of steel or transporting more goods to foreign ports than other nations. China is on track to accomplish the twenty-first-century version of both types of achievements.

The modern equivalent is China's dominance in rare-earth minerals combined with aggressive backing of a space program to help sustain that dominance. And the question it raises is: What will the outcome be this time? Meaning, will China have so much momentum that the dominance will continue through mining operations on Earth as well as the Moon, or will there be an international, space-age equivalent of brute force Western imperialism?

Concern over China's mining dominance with rare earths has been called alarmist by some people, since the "rare" in the name is misleading. The US Geological Survey considers them abundant in the Earth's crust; the problem is they don't necessarily occur in desirable concentrations. In addition, once extracted from the ground, the mixture of rare earths and other elements must undergo complex chemical processes. Depending on how much non-essential material is jumbled together with the rare earths, the steps to extracting the rare earths "may be repeated hundreds or even thousands of times."[5]

Many rare-earth minerals are not only indispensable in making the tools and gadgets of days gone by—rare earths like neodymium and praseodymium play an essential role in the green technology and high-tech industries of the present and future. China dominates rare-earth production, by far, putting out 120,000 metric tons in 2018.[6] Australia is a distant second at 20,000 MT and the United States produces 15,000 MT. Every other country is currently a bit player. China's strength in this arena may make it feasible for the country to leave the World Trade Organization (WTO). If they are in a position to do that, gutting the only global international organization responsible for rule of trade between nations, they might set up an alternative body with rules that are more in line with their economic goals. The result is logical and economically smart for China: It could use its domination of rare earths to keep the price of exports high and the price within China's borders low. That would force companies dependent on rare earths—like the energy and tech companies of the future—to relocate their plants to China.

Are we talking about history or the future here? Both. In 2014, China did try to implement a pricing strategy that would make rare-earth metals cheaper inside the country than outside of it. The WTO put a stop to that. If the WTO became irrelevant, then the past is prologue.

With the United States and Australia leading the world's mini-producers in rare earths, the best question they can ask is, "What do we need to do to stay competitive?" In other words, they would ask, "What can we do to make enough money to remain relevant in the rare-earths economy?"

Some companies like Planetary Resources and Deep Space Industries had plans to extract water and mineral resources from celestial bodies, but those plans ultimately didn't fly as originally conceived. The country of Luxembourg lost its $12 million investment in Planetary Resources, and it was not alone in its loss. Ultimately, both companies were acquired and their plans for conquering asteroids reworked. They weren't wrong in assessing opportunity, just the way Apple wasn't wrong in predicting the massive market desire for a tablet computer. Ultimately, the bugs in the scheme didn't destroy the company technology, but the initial timeline for success could not be realized.

A short-term way of dealing with the need for rare earths until those exploratory efforts get underway is recycling.

Recycling rare-earth elements, once thought to be an impossibility, is now a reality. In April 2019, Apple announced a major expansion of its recycling programs, with a "disassembly robot" named Daisy the star of the new effort.[7] Daisy streamlines recycling processes for the iPhone by taking apart 1.2 million iPhones a year, giving precious metals like cobalt multiple lives. The US government has several agencies exploring rare-earth element recycling as well. Ames Laboratory scientists recycle rare earths from magnets and the US Department of Energy is working on a separation process to make purifying recycled rare-earth elements much less expensive.[8] The lesson in this is, as we continue to push toward opportunities hundreds or thousands of miles above Earth, we need to sustain ambitious innovation on Earth as well.

MISPLACED HUMAN RESOURCES

Many times in the course of human history, men and women have assumed duties for which they were not well suited. As the fields of robotics and artificial intelligence (AI) mature, we will undoubtedly have people doing jobs that should be done by intelligent robots. We have to be smart, not territorial, about some of the exciting work related to space exploration. As much as we might want to do that work with our own hands and see it with our own eyes, sometimes human beings should step aside and yield to the smart machines.

Through the ages it's been evident in the field of medicine, for example, that we have put people in the wrong place, or put the wrong people in positions they should not have held. And as we look toward the Moon and Mars, there is a good chance we could make the same kind of mistake again.

International commerce gave rise to a new profession: ship's surgeon. It was a flawed concept in that qualified surgeons did not tend to seek out the position. Historian Iris Bruijn's dissertation, *Ship's*

Surgeons of the Dutch East India Company, offers this insight: "In sharp contrast to the academically schooled physician, the (ship's) surgeon was largely trained empirically."[9] In the early days of the American colonies, apothecaries had similar on-the-job training, practicing as doctors and, as necessary, performing surgery, according the official website of Colonial Williamsburg.

If that sounds outrageous consider how our most highly skilled physician astronauts would have empirical training once they venture into environments that no human has ever before encountered.

Breakthroughs in telemedicine in the second half of the twentieth century aimed to make healthcare for people in remote locations a methodical, technology-based approach to handling problems when the right specialist could not be present. The intent was to reduce the onsite, trial-and-error of treatment and harness the knowledge of physicians who were located elsewhere.

Some of the pioneering work happened near battlefields, notably during Operation Restore Hope in Somalia during the early 1990s. US physicians deployed to the evacuation hospital outside Mogadishu met with alien threats to health: fine sand, fleas, spitting cobras, three kinds of malaria, and drug addicts armed with AK-47s. They needed expertise from specialists back home to address the ongoing medical surprises affecting American soldiers and allies. That expertise was delivered by a simple system bringing together computing capability, satellite communications, and digital imaging.[10]

One example of how doctors onsite failed to meet a medical need without technological support involved a soldier who developed ugly bumps around his eyes. The ophthalmologist in Somalia made a preliminary diagnosis of lupus erythematosus, an auto-immune disease that would have resulted in having the solider evacuated to an Army hospital in Landstuhl, Germany, 6,000 miles away. Sending an image of the problem to a specialist at Walter Reed Army Medical Center in Washington, DC, averted the evacuation: WRAMC specialists determined the soldier suffered from contact dermatitis caused by an allergy to rubber in the goggles protecting his eyes from the gusts of fine sand.

The rudimentary system used in Somalia was one of many early demonstrations of the effectiveness of remote medicine. Another occurred about two months after the soldier's allergy was diagnosed from 9,000 miles away. This exercise involved a separation of only a few hundred miles, but the distance was straight up.

Bernard Harris's doctor bag was a clear plastic pouch sealed with Velcro. He removed a stethoscope and began the examination. Charlie Precourt, his patient, bobbed away from the doctor with each touch. He would have floated away completely out of range if untethered. As Dr. Harris proceeded, a gold medallion escaped from his shirt and seemed to swim in front of his face. The patient looked relaxed, with a little sway in his back and arms dangling at his side. In fact, Precourt's posture was reminiscent of pictures of prehistoric man. Harris checked one thing after another. He found fluid on the lungs, the heart migrating toward the center of the chest, and other organs shifting and hiding. In short, Precourt was perfectly normal—under the circumstances.[11]

Those circumstances—the first checkup in microgravity—were aboard the space shuttle *Columbia*. The date of the event was May 4, 1993, and it was broadcast from the shuttle to the Mayo Clinic. The intent was to show in real time what normal space physiology is, and to keep the emerging discipline of telemedicine moving forward and upward.

The doctors on the ground at Mayo Clinic saw a quandary take shape—one that has since been at the center of space medicine discussions. They recognized certain causes and effects that ran across disciplines:[12]

CAUSES	EFFECTS
fluid shifts	kidney stones
bone demineralization	fractures
muscle loss	cardiac arrhythmias
electrolyte changes	constipation
cardiac deconditioning	eye problems
closed-loop confined environment	infectious disease skin problems toxic exposures
construction in space and space walks	trauma decompression sickness
accidents	burns toxic exposures radiation
general	space motion sickness

This 1993 clip from history, therefore, implies one of the greatest challenges of putting physicians aboard spacecraft: If we're going to send a human doctor, which kind of doctor should go? Or are we better off relying on astronauts with general medical training with AI having a dominant role—the vision that Jay Sanders ("The Father of Telemedicine") shared with us for this book?[13]

In the early days of telemedicine, one set of humans communicated with another set of humans. We are rapidly approaching a time, however, when humans may have to rely on AI because human medical professionals will be too far away to make a difference. We explore the controversy surrounding man's role more in Chapter 10.

It has not been the mission of the US government to colonize and develop space, even at the height of the Apollo program, but rather to explore it. In addition, no single government could ever develop and settle space—even if it wanted to do so—although single countries such as China and the United States could potentially go it alone as explorers. In areas where we do not want history to repeat itself, we have to embrace high-risk phenomena such as public-private partnerships, intergovernmental cooperation, and man-machine interfaces.

With the wisdom and records of history, we should keep moving forward with the curiosity of people focused on the future. In the words of the fictional Captain Jean-Luc Picard of the Starship *Enterprise*: "There is a way out of every box, a solution to every puzzle; it's just a matter of finding it."[14]

PART II

THE ORGANIZATIONAL
INFRASTRUCTURE

THREE

COMPANIES IN SPACE

D ecades before the birth of NASA in 1958, the United States began awarding cost-plus contracts to stimulate wartime production by American companies. Even large companies could not afford to self-fund the R&D for new technologies sought by the government, so contracting agencies would pay the expenses and a fee so the company could get the technology it needed.

In the early days of space travel, the government cost-plus contract was common. And after President John F. Kennedy's September 12, 1962, "We choose to go the Moon" speech, the push was on to engage the best minds and manufacturing capabilities in the country to support the space race. Performance, quality, and delivery time exceeded the importance of cost—a perfect framework for cost-plus contracts.

Governments still fund a great many space efforts, of course, but the US space industry has been transitioning back to a firm fixed price model. With this model, incremental change can be funded by the government, but it's less likely a revolutionary change will result.

Companies, even small start-ups, have found ways to make technological advances that could be considered revolutionary changes. They have found creative ways to do it both with and without government help.

In Chapter 5, we look at how some of the big transportation companies are managing to innovate and succeed. In this chapter, we look at smaller space companies that have debuted distinctive technological solutions. We pay particular attention to the commonalities and differences in how they faced challenges and developed strategies for profit. Government prime contractors such as Lockheed Martin and Boeing still open doors for smaller companies by inviting them into contracts. Other relatively young companies have managed to supply government agencies directly with innovations. Some pool their financial and technological resources to get their innovations into space.

To help their fellow space entrepreneurs, dozens contributed stories and practical information to this book. Among the topics: financing a new space company, taking it from ideas to tangible results, causes of failure, and catalysts of commercial viability.

FORMING A SPACE COMPANY: ADVANTAGES AND BARRIERS

Space entrepreneurship is a fever we've seen in student populations, within communities of seasoned professionals, and at gatherings of space aficionados. High-profile examples such as Virgin Galactic, SpaceX, and Bigelow Space have captured the attention of both the public and investors. Although these transportation companies are transformational for the space industry, just as significant is the contribution of hundreds of successful entrepreneurial space companies that populate a domain we'll refer to as *eSpace*, a short form of "entrepreneurial space" coined by the Center for Space Entrepreneurship. This not-for-profit organization is a collaboration of industry, government, and academia that aims to spawn and accelerate the development of entrepreneurial space companies and a multidisciplinary workforce to support them.

After selling Starsys, his company that made the space motor for the Mars Pathfinder, Scott Tibbitts became executive director of the Center for Space Entrepreneurship. As part of the Center's efforts to nurture new eSpace companies, he did an analysis of the business environment space companies inhabit—unique in many ways—and he pinpointed both advantages they potentially have and barriers they often seem to face. The original work was done in 2008, and it has largely stood the test of time, but some advantages and barriers have morphed with increased competition, legislative changes in the United States and other countries, refocusing of government agency activities, and the economic impact of the global financial collapse in 2008.[1]

ADVANTAGES

A rather general advantage that many people in the eSpace arena would cite is company mentorship of students, who are then prepared to jump into the deep end of corporate life while they are still in school or immediately after graduation. Twyman Clements, founder of Space Tango, did the groundwork to launch his company while he was still in graduate school, and then partnered with a corporate executive shortly thereafter. (The genesis and growth of that company receives attention in Chapter 6.)

As a corollary, a few entrepreneurs have come into an academic program with an idea of how to build a company around it or expand the scope of an existing company, and the university provides a venue for proof-of-concept. The 3D printer at the Colorado School of Mines that demonstrates how habitats can "grow" from materials found on the lunar surface is a product of ICON, a company founded by Jason Ballard, a graduate student at Mines.[2] In 2018, ICON delivered the first 3D-printed home in the United States to receive a building permit; homebuilding has been the company's core business from the beginning. The 3D printer on site at Mines is a logical application of ICON's technology, potentially expanding the company's business into space. About a story tall and ten feet wide, it looks like sturdy frame with nozzles and wires—something that might be shipped in boxes with "some assembly required" aboard a spacecraft and then set up to go to work on the Moon. While students at Mines do their lab work with turning lunar material into building material, the ICON 3D printer is there for NASA, companies, and funding sources to see the practical results.

Characteristics identified by Tibbitts as traits unique to space industries that support the formation of companies include:

- **R&D funded by customer.** In the eSpace world, the cost for technology development is often borne by the customer, significantly lessening funding required for a startup. Even though the customer, who might be a large company or a government agency, pays the bill, the company retains and develops the intellectual property for future customers. This advantage is not restricted to startups, of course, since large aerospace companies have enjoyed the same kind of R&D investment through government contracts.

- **Government funding sources and incentives.** Space companies, even in the idea stage, have paths to abundant sources of development funding in the United States. The Small Business Innovative Research (SBIR) and Small Business Technology Transfer (STTR) award grants, and Broad Agency

Announcements (BAAs) are ways to secure contracts. Together, SBIR and STTR are known as America's Seed Fund. Startups in many other countries have similar opportunities, including other government-backed incentives for private investment. One example of the former in the UK is the Prince's Trust Grants, which gives small amounts to entrepreneurs who are eighteen to thirty years old. In the latter category is the Seed Enterprise Investment Scheme (SEIS), a British initiative providing substantial tax relief to encourage investment in UK companies. SEIS relief allows investors to claim back up to 78 percent of their investment in the first year, currently up to £150,000 (about $190,000 as of this writing).[3]

- **The SBIR program** is designed to stimulate technological innovation to meet federal R&D needs and it has a specified goal of funding innovations with commercial potential. The program is highly competitive, but there is evidence to suggest that program is becoming more accessible because the number of applicants has dropped. NASA did a study in 2014 looking at the number of SBIR applications in 2005 versus 2014: The number dropped from 1,900 to 942, although the number of awards held steady at about 300.[4] That number exceeded 400 by 2018, so the program should theoretically be attractive to startups with a commercial vision.

- **The STTR program** aims to expand the public/private sector partnership to include joint venture opportunities for small businesses and not-for-profit research institutions. In the STTR environment, the business must collaborate with a research institution. STTR sees its most important role as bridging the gap between performance of basic science and commercializing the innovations that result.

Space-related SBIR or STTR grants can be associated with any number of government agencies or branches of the military, but a search on those given just for NASA projects indicates

that there were 515 awards made in 2018, with most coming
from SBIR (447) and most given for Phase I work (345),
meaning the company has a clear idea of the scientific and technical merits of a new product or concept, and how it can be
commercialized. In other words, they received equity-free
funding because they had a well-articulated concept with de-
monstrable value to the marketplace. Phase II projects always
received larger awards than these, which makes sense since the
companies have already proven the merits of their project in
Phase I. At the same time, the Phase I projects received anywhere from $98,453 to $125,000—enough for a small company with an innovative idea to enter the space industry.[5]

- **BAAs** concern scientific or research projects and some early
stage development efforts. They are notices published online at
the Federal Business Opportunities (FedBizOpps.gov) website
requesting proposals from private firms of any size. Proposals
coming through the BAA process go through peer review or a
similar vetting process and there is a strong emphasis on having
the process reflect full and open competition. When NASA
issued a BAA for a human landing system early in 2019, the
agency went to the trouble of conducting a three-hour industry
forum to give any interested party the chance to hear exactly
what the agency was looking for, and to ask decision-makers
questions. And in a display of full NASA support for the forum, Jim Bridenstine, the Administrator of NASA at the time,
participated in the event.

- **Low-volume requirements.** A typical aerospace contract cou-
ples high engineering content with low manufacturing volume,
which is well suited for startups. A company that can space-
qualify a new technology, probably with the help of an established company or space agency, may only need only a few examples of what it has to offer to get buy-in on manufacturing.

- **Collegial support.** Established space companies often provide a kind of paternal support for eSpace company formation. In their efforts to win big contracts and achieve market dominance, they recognize the value of innovations coming out of startups and how they might provide a competitive advantage.

BARRIERS

The good news is that some of the circumstances creating barriers to entry for eSpace companies have changed for the better. Nonetheless, they still exist and are entangled with each other in some interesting ways—and oddly enough, those entanglements have more upside than downside.

- The aerospace market is shallow and narrow. The numbers of customers and vendors are relatively small. By the mid-twenty-first century when commercial activities in space are common, the landscape will change. For now, the reality of limited short-term growth opportunities and ceilings on profits gives traditional investors reason to step back.
- Lack of professional investment can translate to a lack of professional business expertise for the startup. Throughout the book, there are examples of people with scientific or technical expertise who co-founded companies with a seasoned business executive. At the same time, there are plenty of examples of companies struggling with an inventor at the helm when a business head might have been a better choice in a top leadership role.
- The need for "previous flight heritage" creates a significant barrier to entry. The logical challenge to a new idea is: "What makes you think you understand our problem and the environment where that problem occurs?" ICON's founder and CEO

Jason Ballard demonstrated one way to mitigate it. He is an experienced businessman with an academic background in conservation biology. His company is mainly in the home-building business, an outgrowth of Ballard's long-term concerns about homelessness and sustainable approaches to home construction and improvement. He jumped into the eSpace environment, however, when he enrolled in a master's program in Space Resources at Colorado School of Mines. His "flight heritage" comes from working with Mines faculty as well as executives in space-related companies that rely on the school for expertise and facilities. It has put the company on track to use 3D printing to build habitats on the lunar surface.

- Substantial infrastructure requirements with respect to quality, manufacturing, test, and configuration control are expensive and daunting. We have heard the same message from one space entrepreneur or executive after another: Traditionally, this is a business of perfection. When you send people into space, all systems are expected to function without glitches (although on some unmanned projects the demand for perfection has diminished). There is no escaping the need for the infrastructure requirements that create a barrier to entry. However, the kind of grant support governments provide can help companies through this phase of development. Also, teaming with university research facilities is a way to put keen eyes and fresh talent on a project.

- "It is a cloistered industry with unique government, quality, ITAR, technical requirements difficult to decipher and to penetrate."[6] This is one of the barriers recognized in 2008 that has become less of an issue. According to current eSpace entrepreneurs, there has been an effort on the part of government to become more transparent in technical requirements; moving the need for export clearance from the State Department to the Commerce Department was integral to that change. As a corollary, the International Traffic in Arms (ITAR) regulations were relaxed a bit as nations became more adept at distinguishing between military technologies and space exploration technologies. What political interests are in control of a government can, of course, affect the interpretation of regulations in a nation. Chris McCormick, who built a company called Broad Reach known for very sophisticated high-value avionics, thinks regulations like ITAR should not be viewed as a significant barrier to space companies:

A lot of people complain about ITAR. A lot of things that Broad Reach sold internationally had to go through ITAR, so we had plenty of experience with it. If you want to travel internationally, you'd better go get a passport first or you can't leave the country and the next country won't let you in. Is that good or bad? It's just a process. If you're going to sell internationally, doing the paperwork for ITAR is part of the cost of doing business. The worst horror stories I've heard about ITAR are from people who've never actually done it.[7]

McCormick, who sold Broad Reach to Moog Inc. in 2012 for $46.5 million, is the co-founder and Chairman of PlanetiQ, specializing in cost-effective data for weather forecasting and climate monitoring. He points out that the US government has not only engaged in deregulation affecting space companies, but has also created positions to help them navigate the regulations still in place. For example, the person holding the position of Director of the Office of Space Commerce at the US Department of Commerce, Kevin O'Connell, is tasked with being the official "space industry advocate" in government.[8] His task is to have regulations that both protect government interests and work for the growing commercial space industry.

- The aerospace industry is "relationship-driven" and relationships require time to develop. Again, let's flip this around and acknowledge that a relatively small pool of customers and vendors allows for knowing who the players are and then pursuing relationships. They may take time to capitalize on, but the connections can be made more easily than in many other industries. The challenge is building trust through demonstration of expertise, excellence, and consistent performance; you have to earn membership in the "club."
- Replication involves sourcing talent from a limited pool. However, the pool has grown in size from a bathtub to a lake. Hundreds—if not thousands—of universities worldwide now offer programs in fields directly supporting eSpace activities. The critical factor in areas such as rocket design, however, is that very few offer substantial hands-on experience, that is, launching their designs into space.

In short, there are obviously potentially positive ways of looking at some of the unique circumstances presenting challenges to start ups in eSpace. The limited number of customers and vendors means that you can find out who the key players are in a relatively short time and try to forge relationships with them. With that foundation, it's possible to create the space flight heritage through association with others in the business, at universities, and in government. And with the degrees related to space exploration becoming more diverse and more numerous, talent is rushing into areas such as avionics technology, remote sensing, robotics, and energy management.

COMMONALITIES OF STARTUPS

When Scott Tibbitts did his analysis of eSpace companies, he included the company he had just sold in the mix of those he studied.

In plotting his company's milestones, he realized those pivotal events tracked with many other companies he either knew from personal experience or research.[9] He could see that the commonalities were not inherently positive or negative; they merely illustrated a pattern that might alert other entrepreneurs to signs of impending trouble or victory. Research into eSpace companies that have emerged since his 2008 analysis indicates many of those commonalities still exist. Browsing through the descriptions of startups listed on the Space Bandits website reveals there is a still a common pattern of companies around the world being self-funded by a technologist founder and operating during the startup phase with fewer than ten employees, who are probably contributing sweat equity.[10]

TRAITS AND TIMELINES REPEATED

Summarizing key commonalities that surface in the company profiles below, we could say they generally are:

- Formation by one or more founders with a strong technical background, but limited business experience. Business expertise develops on-the-job and therefore this lack of expertise can be an impediment to growth and success.
- The business is built around a widely recognized industry technological need and/or the recognized competence of the founders in a particular aerospace area.
- Initial funding is required primarily for the founder's time to establish the first contract. This is provided through sweat equity augmented by money from friends and family, which may be crowdsourced.
- Usually minimal start-up capital is involved. Investment is often less than $100,000.

- "Subsistence level" break even can occur early (sometimes in the first year) from a first contract or SBIR.
- The subsistence period continues while the reputation develops.
- A significant catalytic contract triggers a growth period after the subsistence period.
- Growth challenges exist from replication of product, requiring high-touch, high talent.
- A moderate growth rate allows growth capital to be self-generated or banked.

Two of the three companies we chose to profile have decades of history exclusively in space businesses; development shows how the commonalities surfaced and played out over time. One company has only a five-year history so far but it's easy to see many of the same traits and timelines already.

STARSYS: FROM A GARAGE TO MARS

The organization Tibbitts called Starsys Research ultimately succeeded as a company providing mechanical systems for spacecraft.[11] Its pattern of development reflects a common one: from a great idea and sweat equity, he built a company worth millions and saw a dream come true.

Key milestones in Starsys's growth are:

- In 1988, Tibbitts was working in a water heater engineering company. He identified a use for a domestic water heater technology in aerospace, that is, paraffin thermal actuators as a replacement for explosive bolts. It could be said that he then created a space company from candle wax and $7.20 in plumbing store parts.[12]

- The water heater company "incubated" the aerospace entity till break-even—basically a paternal arrangement.
- NASA-JPL provided no funding, but the agency did deliver technical assistance to help Starsys space-qualify the technology.
- The first contract for flight hardware was signed nine months after the company's founding. With it came the space heritage needed to give Starsys legitimacy in the eyes of the industry and government funding sources.
- A mere $10,000 investment from friends and family plus subsistence contracts allowed survival through first year. It was a period when sweat equity kept the doors open.
- Multiple contracts in 1989 made the company self-sufficient, enabling Starsys to leave the incubator and hire core employees.
- Customer advocates encouraged development into spacecraft mechanisms. Starsys then shifted to spacecraft mechanisms as the core product line.
- In 1992, the company hit $3 million in sales and had twenty employees. A transformational $3 million contract catalyzed the growth period for company.
- In 1999, Starsys hit $6 million in sales and had sixty employees. It then acquired a space motor technology company, further catalyzing growth.
- The motor technology capability drove 40 percent growth for the next years to $18 million. By 2005, 150 employees were onboard and the company had broad technological offerings.
- High growth created replication and growth capital challenges, which led to the acquisition by SpaceDev in 2006 to provide additional resources to support growth.

- By 2008, Starsys was a stable $28 million division of
  SpaceDev providing deployable structures, mechanisms,
  and actuators for spacecraft.

In the life of any successful company, there is at least one moment that stands out as a turning point, a moment that justifies saying, "We made it!" For Starsys, that moment was dramatic, just as it has been for myriad other companies whose inventions made it into space; the dramatic instant might even be considered another commonality.

At 9:28 p.m. on January 3, 2004, a Starsys team of eighty-five people had gathered at a Boulder, Colorado, Holiday Inn. Champagne in hand, they were poised to toast a victory—they hoped. Barely breathing, they stared at the screen showing a few tense flight controllers at the Jet Propulsion Laboratory in Pasadena, California. Everyone waited for the same thing: a single beep beamed from a Mars rover 140 million miles away. The beep would only be heard if every one of the dozens of devices made by those people in Boulder worked perfectly after being wrenched away from Earth's gravity and speeding through space for eight months.

In less than a minute, the control room erupted with yells, cheers and hugs. Mission control heard the rover beep. It was "I'm alive!" in rover language. The Starsys hardware had worked, helping enable NASA to land the first mobile science lab on another planet.

Minutes later, the video feed switched to a press conference that opened with Veronica McGregor of the JPL Media Relations Office saying: "Good evening, we're back here at the Jet Propulsion Laboratory in Pasadena, California and we're also at Mars."[13] Pete Theisinger, the manager of the Spirit Rover program, could not stop smiling. He declared: "We are on the surface of Mars!" and acknowledged the great victory as a "team accomplishment."

> We were having some trouble getting some parts built and a company in Boulder, Colorado picked up the challenge to help us out. I went

out to visit them...I was sitting in front of a small building and I said to myself, "If they don't do their job exactly right, we will all fail."[14]

That acknowledgment, heard worldwide, sent the team assembled at the Holiday Inn into an emotional exosphere.

Later that night, Tibbitts went home and opened the NASA website, hoping for any kind of update. The download took a little while, but there it was on his screen: the first picture back from Mars.

It was a surreal moment, seeing that first view from another planet that had been taken just minutes ago, watching along with tens of thousands of people, seeing the first ever picture of bedrock on a planet other than Earth.

After a few minutes I noticed something unusual in the picture. To the left of center, a fuzzy splotch on a spacecraft motor in the foreground. I clicked the cursor a few times, zooming in.

"No way!"

Our company logo came into focus, clearly visible on an image just taken minutes before, 140 million miles from earth.

The next day I discovered that two years prior, one of our designers had researched where the camera would be pointed for the first image and specified where the Starsys logo would be so that it would be center stage two years later, a gift to the company he worked with.[15]

Tibbitts's story hints at another commonality among eSpace companies: a sense of mission and loyalty beyond what's expected.

SEAKR:
CHEAP PARTS + KEEN MINDS = TURNAROUND

A brief history of SEAKR Engineering, a leading provider of advanced avionics and electronics for space applications, illustrates a number of the points made and is consistent with the chronology outlined above.[16] Key additional elements of SEAKR's story are:

1. How success and failure came wrapped in the same package. In an industry as fast-moving technologically and expensive as space exploration, this scenario is not uncommon.
2. How savvy, creative people can turn off-the-shelf parts into space-qualified technology. Their design engineering costs were similar to competitors, but the raw materials cost fell far below that of the competition.

Formed in 1982 by retired Air Force Colonel Ray Anderson and his son Scott, SEAKR's founders were both engineers who had worked for Rockwell International. SBIR had just published its first bulky catalog of funding opportunities. The Andersons studied it and got wide-eyed; the government needed a solid state recorder (SSR)—the first of its kind. The device was meant to replace tape storage with solid state storage on spacecraft. Between Scott Anderson's confidence he could build it and his father's contacts in the industry and government, they determined this was great starting point. After winning the Phase II contract, Ray's son, Eric, joined the team, followed by Kurt in the early 1990s.

Although the first major effort was considered a technological success, and launched the company as a space-qualified memory provider, SEAKR had integrated the wrong technology for future needs—that is, bubble memory.[17] Their reasoning was sound: Bubble memory was extremely radiation hardened, and it was nonvolatile, meaning the power to it could be turned off without loss of data. Technologically, it made sense as a tape-recorder replacement. The downsides did them in. Bubble memory involved a heavy device, it was not user-friendly, and it was not low-cost.

The company survived the subsequent two-year dry period with-out contracts and funding through what President and COO Eric Anderson calls "sweat and love equity."[18] His spouse—also a tech-nical expert who had been employee number four at SEAKR—returned to the workforce outside the company and thereby kept the cash flow at a steady trickle. During that time, the team pivoted from the old approach to a new one, bringing along what they learned from the failure. They sunk their expertise and experience into a technology for future storage: Dynamic Random Access Memory (DRAM).

The dawn of the 1990s marked a time when DRAM gained a great deal of acceptance as a tape recorder replacement. It might be said that SEAKR then pushed ahead of the competition by adopting a technology/fiscal awareness strategy: knowing the team's capabilities, identifying available resources, and knowing how to apply those re-sources to solving customer problems in the most cost-effective man-ner possible.

In 1990, SEAKR won the contract for the Clementine program, a Naval Research Lab experiment to look for water on the Moon. Fortunately for SEAKR, the funding for the project didn't match its lofty goals. Eric remembers, "They didn't have a lot of money. At the time, they coined a phrase for the program, 'Faster, better, cheaper.' They couldn't afford to come to a big player for this type of memory, so they came to us."[19] Delivering the product on time, on budget, and to specifications earned endorsements from the colonel overseeing the project. The colonel's support magnetically drew other customers to the company, ultimately showing the financial value of good will. Contracts for the SeaStar and Apex projects—related to ocean and radiation belt monitoring—soon followed.

At that point, SEAKR was a stable, seventeen-person company with $4 million in revenues. The operations model—and this is something seen over and over among young companies in space businesses—was combining inexpensive parts with a high level of technical expertise. This model also surfaces in the rocket business, with "faster, better, cheaper" being the mantra of companies aimed at sending tourists and other payload into Low Earth Orbit.

SEAKR was able to deliver DRAM storage systems at one-fifth the cost of the competition because they used commercial dynamic RAM. In contrast, competitors used a radiation-hardened static RAM. But wouldn't a government customer planning to send the product into space insist on something that was radiation-hardened, that is, guaranteed not to fail in the face of high energy particles that can make a bit turn from a "1" to a "0"? These high-energy particles are trapped in Earth's radiation belt and cannot be avoided. They will play havoc with electronics sent into space unless protections are in place—an expensive proposition.

The engineers at SEAKR solved the problem a different way, offering the customer a product that ensured data integrity and had about a four-x density advantage over the competition at a fifth of the cost.

- The storage performance came from DRAM's ability to store one bit per transistor, whereas with SRAM, it takes four transistors.
- The cost savings is because a commercial dynamic RAM would have cost about $400, whereas a radiation-hardened static RAM would have cost $2,000.
- The data protection resulted from superior math. An error correcting code is a mathematical way of storing extra bits, of recreating data. It's a technology that had been used with disk drives, but SEAKR applied it to DRAM. Powerful error correcting codes would kick in to recreate any damaged data.

At that point, SEAKR could claim strong industry relationships and technological excellence. The company was awarded a transformational $40 million contract by Lockheed Martin and the founders moved SEAKR and twenty of its employees to Denver to support the contract in 1995. The contract catalyzed industry acceptance. Two more major contracts followed on the heels of that one, enabling SEAKR to take the lead in the SSR market. Competitors had three elements going against them from that point on: they fell behind technologically, became less relevant as a result, and therefore missed contract opportunities.

For SEAKR, growth and diversification followed. Seeing around the corner, they turned their attention to computer processing systems for satellites. As of 2019, the company had 450 employees and $130 million in yearly revenues.

OMS: WEATHER DATA WITH TACTICAL VALUE

Brian Sanders was a very shy child. His first-grade teacher tried to nurture his people skills by asking him to give a weather report to the class every day. The assignment energized him. He would get up a five o'clock with his father and catch the early morning weather report. On his birthday, he even visited his local television station and watched the weather report live in the studio.[20]

As a co-founder of Orbital Micro Systems (OMS), weather became an integral part of his career. OMS aims to provide access to realtime, actionable data about weather and ground conditions to customers to whom logistics considerations are critical. Microwave remote sensing technology combines with a data system that enables timely distribution of searchable information to customers. Prime target industries include agriculture, aviation, disaster response, finance, energy, insurance, and maritime.

The company was started by a team out of the University of Colorado, Boulder, who knew previous missions with hefty bottom lines and proprietary instruments were not delivering the weather data required by many organizations. The knowledge base would probably continue to diminish without another round of government investment in launching huge, sophisticated weather satellites. And then sometime in 2014, it occurred to them: Maybe we can miniaturize the technology—specifically, use the small satellite concept and widespread distribution to effectively democratize weather information nationally and globally. Their exposure to CubeSat projects, which were thriving in many university labs, helped trigger the epiphany. CubeSats are small cubes containing research projects done in space. NASA's original intent was to help academic institutions develop space programs by lowering the barriers for using the ISS for complex science and manufacturing.

While at CU Boulder, Sanders was the Deputy Director of the Colorado Space Grant Consortium, part of NASA's STEM (science, technology, education, and math) education division enabling students to get real-world, hands-on flight experience. In forming OMS, he joined with world-renowned experts in microwave radiometery, NASA data distribution systems, and others with key skills to start an Earth-observing satellite company.

Like many space companies emerging in the current environment of commercialization of space, they took the fundamental ingredients of technological expertise in the critical fields and mixed them with a strong grasp of market need. And like many companies before them that have tackled complex technological issues, OMS needed about three years before they could produce observations from space. OMS launched its first satellite—about the size of a shoebox—in April 2019 and has been getting operational data from it since shortly after that time.

Let's jump back in time to the 1970s, when remote sensing scientists at NASA's Goddard Space Flight Center conducted studies based on Landsat data. Even before that, satellites were taking pictures of our world, but the Landsat program set precedents in delivering data usable to governments and industries on Earth. A joint program of NASA and the US Geological Survey, it is the longest continuous space-based record of Earth's land in existence.[21]

With Landsat, scientists on Earth could see weather and its effects on a global scale. It was valuable—without a doubt—but it was far from realtime even decades into its development. The philosophical underpinning of OMS that drove R&D was the desire to provide a low-cost solution to providing Landsat-like data affordably and without discrimination.

With the OMS technology, they could see through clouds. They could pull out 3D temperature profiles or humidity profiles of the atmosphere—and focus deliver of data on how that information impacts people and their livelihoods. Sanders notes,

> So much of what we see in space companies is this same kind of democratic thrust: Make the technology as accessible and affordable as possible. At the same time, this shouldn't be characterized as altruistic. New space companies are meeting market demand in a way that government-sponsored activities never could—and were never intended to meet.
>
> As extraordinary, comprehensive and powerful as Landsat data is, the question is: How do we make that huge bucket of bits valuable to an end-use customer? That's what OMS is trying to do differently— to serve as a repository for the data collection by OMS and consolidating with other related data such as that from Landsat. The aim is to fuse all of into a common gridded data product.[22]

In the commercialization of space exploration, bypassing bureaucracy as much as possible clearly has found a place in the business plans of eSpace companies. Here is a use case illustrating how a private company like OMS can do it: If a farmer plants seeds but the soil temperature is not conducive to germination, there could be a dramatic reduction in yield due to limited germination or even a catastrophic loss. Before it's too late to replant, the farmer's insurance company might contact him and say, "We have data that indicates improvement in conditions; replant now." The value of that kind of forward-looking tactical information for multiple industries is enormous. The B2B relationship between a company like OMS and the insurance company has a track record of being more efficient in delivering real-time critical data than G2B (government to business) communication.

Like the other companies profiled in this section on commonalities, OMS has also benefited from the SBIR program. Sanders says, "SBIR grants are hugely enabling for small companies like ours."[23]

THE STANDARD FOR WINNING IN THE INDUSTRY

Chris McCormick has had a guiding principle throughout his decades of success with space companies. He passes it along to other entrepreneurs in the form of a provocative question: If your product is sitting on a shelf alongside other competitive products, and every one of them was free, why would the customer choose yours? When you know the answer to that because of your deep knowledge of the competition and your certainty about the excellence of your own product, you should win the customer. In contrast, when you believe your own hype and your company leads with hype instead of genuine competitive advantage, then go back to the drawing board.

> People often lead with money. I'm going to make this or I'm going to do that and make three times my investment. They need to talk about what they want to get done. They should be asking themselves, "Why would anybody buy it? What is the capability needed by the customer? Why does what you are doing matter?"[24]

INFRASTRUCTURE:
GOVERNMENT VISION, INDUSTRY IMPLEMENTATION

Regardless of the size of the company, we have to take an all-hands approach that will make the vision of space exploration and settlement a reality sooner rather than later.

As humans move into space with commercial goals in mind, we will automatically have some people who start to think of themselves as inhabitants of space rather than visitors to it. In some ways, they will be like the people who pushed west in the early days of the United States and built homesteads in Indiana, Ohio, and eventually every other place from the East Coast to California. Along the way, they created the infrastructure to enable themselves to live safely and productively in new environments. In other ways, they are vastly different from the adventurers who responded to the imperative, "Go West!"

The development of space infrastructure will necessarily be different from pioneering eras on Earth. By the time space settlers get to where they are going, they will already need to have critical elements of infrastructure in place or ready to deploy: connectivity, traffic management systems, and habitats, for example.

In terms of vision, governments might be leading the way in looking at how infrastructure might be built. In practical terms, however, meeting these fundamental needs is what industry will do. Practically speaking, this is not a flawed approach; it's a progressive and efficient one. But it could complicate how components of infrastructure are managed and accessed.

People share infrastructure. And when a deer crosses a highway, the deer shares infrastructure as well. We face challenges in sustaining that Earthbound model in space, however, where the infrastructure will likely be built primarily by investment money rather than tax revenue.

The challenge is not without historical precedent. In 1970, Union Oil Company of California and Unocal Corporation discovered oil off the coast of Indonesia. Called the Attaka field, the location was twelve miles out in the Macassar Strait. The nearest land was Borneo, specifically the Indonesian portion of the island called Kalimantan. No infrastructure existed—no roads, buildings, or equipment. Attaka field had only one thing going for it: at least 400 million recoverable barrels of oil.[25] Over the next seven years, Union Oil spent $500 million developing Attaka, as well as four other Kalimantan sites where the company had located oil. Despite constraints put on Unocal by the Indonesian government—foreign companies could only be operators of such sites, not owners, so control was limited—the company generated massive profits through the years. Its current owner, Chevron, is still one of Indonesia's largest oil producers. The fact that the company could build infrastructure but not have proprietary interest in it led to continued development in Kalimantan by both private enterprise and the government.

The bottom line on this is about the bottom line of participating companies: Create a profit incentive, combine that with the mission to contribute to space exploration and settlement, and the result can be an infrastructure supported by governments that supports personal safety, efficient travel, and good work and living conditions in off-Earth environments.

FOUR

GOVERNMENTS IN SPACE

The Goodwill Moon Rocks symbolize how governments could interact if all nations share the adventure of space and reap the rewards of exploration. It's a big "if," but at least humankind tried to start out with positive intentions. Cut from a 2.95-kilogram (6.5-pound) chunk picked up by *Apollo 17* astronaut Harrison Schmitt, the rocks were given to heads of state of 135 nations, to each of the United States, and to each US territory. Astronaut Eugene Cernan, who was on that December 1972 moonwalk with Schmitt, described why the they chose that particular sample to bring back to Earth:

> It's a rock composed of many fragments, of many sizes, and many shapes, probably from all parts of the Moon, perhaps billions of years old. But fragments of all sizes and shapes—and even colors—that have grown together to become a cohesive rock, outlasting the nature of space, sort of living together in a very coherent, very peaceful manner. When we return this rock or some of the others like it to Houston, we'd like to share a piece of this rock with so many of the countries throughout the world. We hope that this will be a symbol of what our feelings are, what the feelings of the Apollo Program are, and a symbol of mankind: that we can live in peace and harmony in the future.[1]

Cernan's words captured the utopian sentiment behind Gene Roddenberry's *Star Trek*, however, governments of the world generally did not follow the script.

GOVERNANCE

Government in space remains a very Earthbound discussion since we are years away from having settlements on the Moon or Mars. The key issue dividing opinions on how governance takes shape is "who owns what?" The governments of Earth may have largely agreed not to plant flags everywhere they land on celestial bodies, but companies from different nations want the right to plant "corporate flags," meaning they have some ability to own and control what they find. Commercialization of space therefore has profound ramifications when it comes to governance.

In the first of the *Federalist Papers*, written by Alexander Hamilton and published in 1787, this statesman and scholar wondered "whether societies of men are really capable or not of establishing good government from reflection and choice, or whether they are forever destined to depend for their political constitutions on accident and force."[2] Since then, humans have proven they are capable of reflection and choice, although with space activities now moving at a rapid pace, the incursion of accident and force in shaping governance in space is probable. Although Hamilton wouldn't approve, what we probably will get is influences from both ends of the spectrum. What has been done through United Nations initiatives and treaties is a good proactive foundation. The "accidents" of discoveries in space and the "force" of corporate pioneers will no doubt involve gamechanging reactions to rapid change.

Years before United States astronauts planted the Stars and Stripes on the Moon, President Lyndon Johnson faced a diplomatic conundrum: Does the country that lands on the Moon first get to claim it? That would be fine if the United States could beat the Soviet Union there, but there was no guarantee the Soviets wouldn't win the race. And what if several countries got to the Moon and tried to set up weaponry, targeting their earthly enemies? Or steal something of heavenly value?

In September 1966, television viewers first heard about the United Federation of Planets during the first *Star Trek* episodes. "The Federation," as it was commonly called, refers to an international and interstellar government supporting peaceful, cooperative exploration of space. Three months after the debut of the Federation on TV, the United Nations General Assembly endorsed the Outer Space Treaty, formally "Treaty on Principles Governing the Activities of States in the Exploration and Use of Outer Space, Including the Moon and Other Celestial Bodies." They opened it for signatures a few weeks later on January 27, 1967. The United States, Soviet Union, and United Kingdom led the way in getting countries to sign the treaty, which prohibits placing nuclear arms or other weapons of mass destruction in orbit, on the Moon, and on other bodies in space. The treaty also says that no nation can claim the Moon or other celestial bodies as its own.

The mid-1960s were tumultuous years of conflicts in many countries, but they were also a time when peaceful coexistence and a multicultural approach to governance became thematically important in both sci-fi entertainment and the real space race.

As of 2019, 132 countries had signed the Outer Space Treaty. In theory, it seemed the dreams of Trekkies everywhere might be realized: International relationships in space would be a lot more harmonious than on Earth. One big reason is that no country could own what it found. That point had so much importance to the creators of the Treaty that Article I of the document explicitly lays out the rules of sharing:

> The exploration and use of outer space, including the Moon and other celestial bodies, shall be carried out for the benefit and in the interests of all countries, irrespective of their degree of economic or scientific development, and shall be the province of all mankind.
>
> Outer space, including the Moon and other celestial bodies, shall be free for exploration and use by all States without discrimination of any kind, on a basis of equality and in accordance with international law, and there shall be free access to all areas of celestial bodies.
>
> There shall be freedom of scientific investigation in outer space, including the Moon and other celestial bodies, and States shall facilitate and encourage international cooperation in such investigation.[3]

The composition of asteroids and the lunar surface used to excite scientists, but as word got around that they contain an abundance of precious metals—gold, silver, platinum—entrepreneurs began itching to go there like the prospectors who raced to California in the mid-1800s. Realizing that asteroids also have iron group metals, other entrepreneurs realized that these rocks found in Low Earth Orbit bodies would be a handy source of essential materials for LEO construction projects, as well as building facilities on the Moon.

What happened next was inevitable—and fueled both excitement and confusion in terms of accessing the resources on celestial bodies.

In 2015, the United States Congress passed the US Commercial Space Launch Competitiveness Act, which had the support of both the Republican and Democratic parties. While the Treaty opened with a statement of sharing and cooperation, this law has language near the top with a statement of purpose that implies profit and competition. Its proclaimed purpose is to

1. facilitate commercial exploration for and commercial recovery of space resources by United States citizens;
2. discourage government barriers to the development in the United States of economically viable, safe, and stable industries for commercial exploration for and commercial recovery of space resources in manners consistent with the international obligations of the United States; and
3. promote the right of United States citizens to engage in commercial exploration for and commercial recovery of space resources free from harmful interference, in accordance with the international obligations of the United States and subject to authorization and continuing supervision by the Federal Government.[4]

The United Arab Emirates and Luxembourg also enacted laws approving private mining of celestial bodies. A number of countries have condemned these laws, citing the 1979 Moon Agreement, an outgrowth of the 1967 Outer Space Treaty. The Agreement explicitly prohibits commercial exploitation of Moon resources, although it's all gums and no teeth. Only six nations signed it, not including the Russian Federation, the People's Republic of China, or the United States.

Regardless of whatever well-intentioned people had in mind in getting international buy-in on treaties dictating a wealth-sharing model, the reality is that space exploration is expensive, and governments are not able or willing to fund it solely with tax dollars. And then there is the resentment factor. The United States funds NASA with about one-half of 1 percent of its annual tax revenue, roughly $20 billion annually. If Somalia were giving as much to a space program, translated into US dollars, that amount would be about $7 million. Somalia has no space program, however; at this time, it doesn't even have an infrastructure in which one could develop. Yet, as one of the world's poorest countries, it would be in a position to make economic strides forward in the idealistic, wealth-sharing model.

The reality is that private sector leadership, with government agencies as partners, is required to solve the financial challenges of space exploration. If control of space activities rests with the world's governments, we will be stuck on Earth forever. The controversial 2015 US law is simply a reflection of reality: Exploration of space has potentially huge benefits for humankind, but we won't have the cash do it without giving private enterprise a profit incentive.

The conundrum of governance, therefore, is how to fuse the ideals of parity and sharing with the practical necessity of rewarding corporate, risk-taking adventurers for their discoveries and efforts.

Those corporate adventurers will also bring human pioneers with a desire to benefit personally from their risk-taking. They will build a settlement on the Moon and ultimately organize themselves into a body politic. Others will build another settlement with their own organization and rules for belonging.

When settlements like this exist, will the settlers be self-governing and pay taxes that stay on the Moon for infrastructure investment? Or will their respective governments get a piece of their earnings because they remain part of the citizenry? Will the United Nations oversee the funds? Or will their status as Moon-dwellers mean they don't pay taxes at all—the companies dominant on the Moon will be responsible for infrastructure related to safety, transportation, power, and buildings? These questions related to a central governance issue have been pondered by the diplomatic community in the proactive spirit of "reflection and choice." At the same time, they roil curiosity about the role that unpredictable developments will ultimately play.

The extent to which countries can collaborate on space activities is one indicator of how much "reflection and choice" will guide space governance.

PARITY VERSUS PREEMINENCE

In many practical ways, governments have demonstrated an extraordinary ability to cooperate on space missions. A prime example is that the space agencies of Canada, Europe, Japan, Russia, and the United States have collaborated in a politically complicated program to sustain the International Space Station. They had ample reason to do so. The continuing and emerging motivations behind that level of cooperation come down to four main ones, according to a study done by the Center for Strategic and International Studies. And even though the authors presented their findings in 2006, the good sense of their conclusions endures, as does their insight into the challenges to that good sense.

The authors posed the question "Why do nations choose to cooperate in space exploration?"[5] and they answered it with these reasons:

- Saves money.
- Generates diplomatic prestige.
- Increases political sustainability.
- Enables workforce stability.

On the surface, all of these sound like practical reasons for sustaining the spirit of the Goodwill Moon Rocks. A close look, however, suggests that any nation's desire for preeminence blows holes in the reasons.

SAVES MONEY

Countries financially unable to meet their space exploration goals without international cooperation savor the idea of partnerships that save money. And then there are the two elephants on the planet: The United States and China could probably each bear the cost of space exploration without partnering with other nations. The snag in a plan for space dominance would be their reliance on private industry. As long as there is a global economy, the companies required to carry out missions would not be supplying equipment and systems that are purely American or Chinese. In spite of the global economy, however, both the United States and China have taken measures to mitigate cross-contamination. In the name of national security, both countries restrict exports of sensitive technology to the other country—a topic explored later in this chapter. Saving money on space exploration is not as important as military and technological superiority on Earth and in the heavens.

GENERATES DIPLOMATIC PRESTIGE

Luxembourg's aggressive moves into space illustrate how a country can use space exploration as a diplomatic tool to elevate its stature in the world. The country is about the size of the state of Rhode Island, the smallest of the United States; to look at it another way, the United States is 3,803 times bigger than Luxembourg. Yet the country is set to sign a memorandum of agreement with NASA to do joint development of space activities. In addition, Luxembourg invested (and lost) €12 million ($13.2 million as of this writing) in Planetary Resources, a US company founded to mine asteroids. The involvement of the two countries sends a signal to the world about strong US-Luxembourg relations—but any diplomatic prestige to be gained would belong to Luxembourg. The small European nation bounded by Belgium, France, and Germany needs the link to US government and commercial space activity more than the United States needs Luxembourg.

Luxembourg has taken other actions to raise its profile and strengthen relationships with the space community worldwide. For one, it established a hospitable legal environment for space companies to base operations there. Primarily, the law gives companies with a Luxembourg presence the right to own and commercialize whatever they do in space. The US counterpart applies only to majority US-owned companies. Whether this law withstands challenges from the international community—some argue both the Luxembourg and US laws run counter to the 1967 Outer Space Treaty—is still in question. Signs point to Luxembourg making more friends than enemies with it, however, so the diplomatic prestige may keep increasing.

INCREASES POLITICAL SUSTAINABILITY

When governments choose to cooperate on space activities, such as the ISS, they enjoy the perception of trust and reliability—at least in one area. The cooperating nations created a partnership with sufficient demonstrated value to be continued through financial difficulties and conflicting agendas in other areas. There is a hopeful aspect to the arrangement: Even quasi-enemies can manage to sustain a productive relationship if it meets common goals. That can be a big "if" for a head of state that sees diminished utility in the arrangement for his or her country. Political sustainability might seem less important or rewarding than withdrawing. The underlying rationale is, "Why cooperate when you might be able to dominate?"

ENABLES WORKFORCE STABILITY

Traditionally, people in a country like the United States who are already employed in space businesses would probably see the stability of employment strengthened by international cooperation. The more that cooperation means relaxations on exports and imports, however, the more likely certain members of the workforce will see an erosion of stability due to outsourcing. Another consideration is the incentives countries can create, and are creating, to make relocation desirable. In short, workforce stability is one of those reasons to cooperate identified by CSIS that has been undercut by trends in the global economy.

PUBLIC AND PRIVATE LAW

International collaboration on space activities has fostered some commonality in thinking and proactive work in developing international space law. As we point out in the governance section, however, commercialization of space has added a reactive quality to many nations' attempts to set policies and establish law.

In 1988, Iowa State University Press published *American Space Law*. Author Nathan C. Goldman was an adjunct professor of space law at the University of Houston and had written another book called *Space Commerce: Free Enterprise on the High Frontier*. Private entrepreneurship clearly held great appeal for him.

Goldman, who divides his book equally between international space law and domestic, broke out with fresh ideas in his book, many of which have withstood the test of time. Unlike others that focused on Earth laws meant to exert influence over space activities, his book moves from those to the private and domestic laws of space. Public laws relate to challenges such as government agency approval for medicines manufactured in space—that is, public concerns. Private law covers issues such as antitrust, insurance, and contracts.

He begins by asserting there are two phases of space law. The first lasted from 1957 to 1980—from Sputnik to the Space Shuttle—and he calls it "the era of international public law"[6] in which the United States and Soviet Union competed for dominance. It was followed by the second phase, when the movement toward commercial development of space began taking shape as did a call for multinational access.

Goldman then makes a provocative assertion—one that should be debated as we look at the role of government in space.

> In the space law realm the new era can be categorized by the atrophy and obsolescence of most space treaties and the shift toward a domestic law of outer space. Under this domestic law private enterprise and other actors have begun to enter space, answering chiefly to the law of their own nation states.[7]

The spirit of those atrophied and obsolescent treaties was one of cooperation and shared opportunities. These treaties may sound naive in terms of practical risk management, intellectual property, and antitrust laws—all examples of so-called private laws—but as governments continue to evolve their role in space activities, the existing treaties shouldn't be dismissed as irrelevant documents. Nor should the needs addressed by the so-called public laws, which aim to establish not just an international public legal order, but an interspatial one.

Keeping in mind that Goldman published his ideas more than three decades ago, we are looking at his important work as a springboard for covering the kinds of legal and regulatory activities that governments might conduct now and in the near future—and which ones they should probably stay away from and leave to private enterprise to sort out. National security, of course, is one area where a government presence will be felt, at least for the foreseeable future.

THE IMPACT OF NATIONAL SECURITY CONCERNS

Through the years, space commerce has faced many roadblocks to development in the name of national security. The intentions are understandable, but the implementation by governments can sometimes be defective. One example is the Missile Technology Control Regime (MTCR), which was established in 1987 "to limit the spread of ballistic missiles and other unmanned delivery systems that could be used for chemical, biological, and nuclear attacks."[8] The thirty-five members of MTCR are urged, but not required, to restrict their exports of certain types of missiles. With advancements in rocket technology, it's easy to see how some twenty-first-century companies could be crippled by overzealous bureaucrats.

National security concerns shaped two US initiatives in 2018 and 2019 that illustrate how one superpower's actions can affect opportunities in space in the very near future; they are explored below. Both have champions, that is, people who feel they are moves to prevent erosion of US economic and military strength. Both have critics, that is, people who feel they are either ill-timed or ill-conceived.

The short-term effects are quite real in terms of job creation and job destruction whether these are initiatives remain intact, morph, or are killed. Before exploring their impact, consider that history has taught us great deal about the kinds of surprises, good and bad, that can come from government initiatives, as well as the lack of them. History has also taught us that much of what is proposed by politicians is more about branding or re-branding than actual change.

On January 10, 1941, President Franklin D. Roosevelt asked Congress to abandon isolationism and support the lend-lease program to provide military aid to countries whose strength was vital to US security. Isolationism had been a central tenet of 1930s policy and it frightened some people and confused many others that the United States was now going to send dollars and equipment to other countries to aid their defenses. Lend-lease appropriations aroused fierce criticism for months. And then Japan bombed Pearl Harbor and the majority of Americans shed their isolationist skin. Lend-lease had strengthened ties to, and defense capabilities of, the very people we would now share the trenches with. During the lend-lease program, which ended four years after it began, the United States provided nearly $50 billion in equipment, materiel, and other kinds of aid to Allies—roughly $700 billion in 2018 dollars. The program was considered a rousing success, with the US ultimately reaping both economic and political benefits.

On the flip side, a *lack* of government initiatives caused the loss of 8.7 million jobs in the United States due to the collapse of the banking system in 2008. When government involvement was truly needed, it was ineffectual at best.

Many government initiatives reflect the perceived needs of the time, but they probably won't last forever, at least not without some amendment. At the same time, even initiatives with short life can provoke dramatic economic changes. Regardless of how aggressively private companies are moving into space businesses, legislative and regulatory actions of governments will make a difference to their success. Conversely, if government does not take responsibility for oversight when companies run amok, the repercussions can be devastating.

The Foreign Investment Risk Review Modernization Act (FIRRMA) is a 2018 law that has caused some anxiety in the commercial space community. It expanded the role of the Committee on Foreign Investment in the United States (CFIUS, pronounced SIFF-ee-yus), an interagency panel tasked with examining investments with potential impact on national security. CFIUS reviews under the new law have scuttled or delayed some major foreign investments in US space companies; at the same time, other reviews have created no problems at all. A $426 million investment from a Chinese company in the satellite company Global Eagle was terminated as a result of the review. On the other hand, a merger of a Canadian and US space company valued at $2.4 billion went through after only minimal difficulty.[9] Implementation of the 2018 act may eventually become a smooth process. In the meantime, however, with space ventures sprouting up all over the planet, encumbrances to investment for even a few years could abort competitive advantages for some American companies.

Another initiative—this one high-profile—could be a rebranding exercise or it could be a real programmatic shift. Either way, it offers a new career track for men and women drawn to military service.

In February 2019, President Donald Trump signed the directive centralizing military space functions under what he called the "space force." Although it would no doubt create a viable career track within the military for space professionals, some see it as a rush to build one part of the space infrastructure without attention to others. Others see it as nothing more than a rebranding of the Air Force Space Command with the addition of personnel with "space" in their title like "Undersecretary of the Air Force for Space." The proposed rollout would have the space force up to full operational strength by 2024.

A space force is not inherently a terrible idea, particularly because we can expect some threat to US companies and citizens as exploration and commercial development of near-Earth bodies occurs. But if the creation of it is more than a rebranding exercise, this initiative will inflate bureaucratic budgets and could be a step toward weaponizing space. Other government initiatives could help meet needs that are both humane and imminent in term of exploration and development.

BAD DRIVERS AND BANDITS

Whether through a Space Force or the current US Space Command or some other government entity, the United States is trying to establish a leadership role in critical infrastructure challenges in space activities. Traffic management and law enforcement are certainly areas where government bodies have a role. The issue is how to make that government presence the embodiment of international expertise and priorities.

With SpaceX alone launching a projected 42,000 satellites for its Starlink project to provide ubiquitous internet access, the number of objects in the sky keeps moving upward quickly. Traffic issues could be mitigated if automatic collision avoidance software always worked, but one of SpaceX's first sixty satellites suffered from a computer bug so it nearly collided with the European Space Agency's Aeolus satellite in September 2019.[10]

Currently, cooperation among space companies and government entities has given rise to a loose traffic control plan, even though space traffic management is a responsibility of the US Space Command. Satellite operators use space surveillance data gathered by the US Air Force to pinpoint the location of their spacecraft. They complement this by sharing tracking information with other operators. If the Air Force and the operators determine that collision seems likely, the operators confer and decide which satellite should move. After the near collision of the SpaceX craft with ESA's, it became apparent that space traffic rules and communication protocols needed standardization and oversight. The International Organization for Standardization (ISO) already has more than two hundred space-related standards on the books that were developed by participating technologists from around the world.[11] This may be a model to move forward, at least in the realm of communications.

US Space Command wants to address the problem by focusing on their military duties in space and handing over responsibility for traffic management to the US Department of Commerce or Transportation. They aren't particularly concerned which entity gets the job, as long as it leaves their purview.[12] The final section of this chapter looks at a proposed alternative, but even that is a US-controlled approach. Other countries will probably demand a more egalitarian system of managing traffic in LEO and outer space.

Turning to crime suppression and law enforcement, the challenge in space will be how very differently space-minded countries handle these challenges on Earth. Would police and military personnel serving a law enforcement function be armed? The Outer Space Treaty prohibits all weapons, including sidearms. In the United Kingdom, Norway, Ireland, Iceland, and New Zealand, police officers generally do not carry guns; their military personnel do, of course. Other countries have police armed with guns of widely varying capabilities. Americans would assume that military personnel, such as members of a Space Force, would have access to powerful weapons, whether or not they carry weapons at all times. Harmonizing law enforcement—and peacekeeping—efforts among countries actively involved in space commerce could be one of the most challenging governance issues.

GOOD JOBS FOR BUREAUCRATS

Madhu Thangavelu, renowned for his work in space architecture, articulated critical roles for government in space in a 2012 article for *Space News*. In proposing a US Department of Space be added to the presidential cabinet (he isn't the first to propose this, but he is a dominant voice), Thangavelu offered ways it would benefit commercial development.

First of all, Thangavelu sees the role of the department as critical in developing infrastructure, particularly "space solar power, orbital debris mitigation, fuel depots, interplanetary missions, or even large space based observatories," which he says should remain the domain of NASA and the government.[13] He does note, however, that servicing such systems could fall to the private sector.

Other department responsibilities would include leading efforts to:

- Align the projects and goals of various spacefaring nations and assist in global projects such as international crewed missions to Mars and other celestial bodies.
- Help coordinate the activities of fledging private space companies.
- Establish a rule of law, with proper law enforcement activity. This is simply to bring order to the space environment, ensure that human life and property is respected and protected, and to enforce the laws already on the books.
- Lease space, not control. This means a parceling of land, filing claims, and financial protection. This is vital because in the past, private space companies have proposed projects that have been squashed by NASA to protect its charter and monopoly. MirCorp is the classic example. With a Department of Space, future proposals like this could be supported, but with NASA having little of no say in the matter.

The department would not threaten NASA's existence, however. NASA, returning to its original charter of R&D, would be a branch of this department. Thangavelu and his University of Southern California team researching this project suggest that the department should have an annual funding starting at about $60 billion a year, consistent with the funding of other departments. NASA would receive $20 billion of this funding for all their research projects, about 15 percent less than its current budget. The remaining $40 billion would be for the department to handle all coordinating functions among large global infrastructure development projects, other partner agencies, and the private sector.

PART II

THE TECHNICAL
INFRASTRUCTURE

FIVE

TRANSPORTATION, FROM HERE TO THERE AND WHEN WE'RE THERE

n 2007, Del Carbon made three flights. One of them made it all the way 21,000 feet. Named for its carbon-fiber construction, and with a nod to the student team's favorite fast-food restaurant (Del Taco), it was the first rocket designed, built, and launched by University of Southern California's Rocket Propulsion Laboratory. As of 2019, another nine rockets had been launched, with the last in the series making history.

> On its fifth space attempt, USCRPL passed the Kármán line, which sits at 330,000 feet above sea level, and fulfilled its founding mission to be the first student team to put a student designed, built, and operated rocket into space....The successful rocket reached 339,800 feet.[1]

For two generations before USCRPL made it to space, governments had a monopoly on enabling space travel. During a 1944 test flight, a German V2 rocket was the first to cross the Kármán line. Thirteen years later, the Soviets blew past it with *Sputnik*. NASA's early entry into spaceflight starred Ham the Chimp, whose historic trip began on January 31, 1961. Finally, in 2004, *SpaceShipOne* became the first privately funded spacecraft to cross the Kármán line, winning the $10 million Ansari X Prize. The spacecraft was the product of a joint venture between Microsoft co-founder Paul Allen and Scaled Composites, founded by SpaceShipOne's designer, Burt Rutan.

Now, the skies are full of privately funded spacecraft, both cargo transport and crew transport vehicles, and space agencies are looking to buy seats on board. Part of the payload will also be ground transportation designed and manufactured by companies for the lunar surface. Gwynne Shotwell, president of Space X, has gotten significant coverage for her confident description of the Falcon Heavy rocket's ability to launch into orbit any payload that has been conceived of to date.[2] Getting off the ground and carrying equipment and people to the Moon or Mars are just pieces of the space transportation picture, however. In the process of going to Earth orbit, the Moon, the asteroids, and beyond, there must be an infrastructure that eventually includes way stations orbiting both Earth and the Moon, with refueling facilities; the equivalent of space taxis; and other factors that private industries cannot afford on their own. Reliance on cooperation is inherent in building the technical infrastructure for space commerce and travel.

SpaceX, Virgin Galactic, Blue Origin, and the other pioneers of providing low-cost access to space will be working side by side in some kind of coordinated fashion, regardless of their status as competitors. No doubt, they will also engage with (or acquire) smaller innovators in avionics, propulsion, and other areas vital to the transportation effort.

BUILDING A BETTER ROCKET

We opened the chapter with the stunning triumph of USC students and *Traveler IV*. What anyone interested in a space career should also know is that the stunning failure preceding it laid the groundwork for their success the following year. *Traveler III* suffered from a miscommunication error, which launched it without the avionics unit being armed. Since that unit was needed to make the recovery system operational, the rocket slammed into the ground, turning into hundreds of metal fragments.

As much as space professionals talk about the desire for perfection, we all know that "perfection" tends to come after imperfections are systematically identified and corrected. Crash-and-burn landings and explosions are part of the learning curve. India's loss of *Chandrayaan-2*'s Vikram moon lander in September 2019 due to a failure of its braking rockets on touchdown not only helped India with "next time" but also fed the database of every country and company aiming for the Moon.[3]

"Perfection" is also not as integral to thinking as it had been when Alan Shepard Jr. became the first American in space on May 5, 1961. Appreciation for lower-cost launches and redundancy in satellite data, for example, allows a lot of less high technology to reach space. In spending less for the launch, the reasoning is that one can afford to send more hardware. If a company, or a government, is spending a billion dollars to put an experiment in space, there is a huge amount of diligence about whether it's worthwhile, likely to succeed, able to inform other work, and has validated merits scientifically and commercially. If the company or government is spending only $40 million to do that same thing, a certain failure rate is tolerated, or even expected. That said, when it comes to rockets, "perfection" is still the aim with the alternative potentially being a devastating explosion.

In the race to make a better rocket, engineers look at what kind of engine it has. With tankage being the most limiting factor when it comes to vehicle design, what goes in the tanks is a key factor in how powerful the engine is and how far it can go.[4] The contents of the tanks also affect how something might go wrong. SpaceX's Falcon 9 rocket exploded in September 2016 because of a problem with a pressure vessel in the second-stage liquid oxygen tank.[5]

CHEMICAL PROPULSION

Falcon 9 was powered by a Merlin engine, a workhorse for early SpaceX launches that is kerosene-based, with liquid oxygen as the oxidizer. This is an amalgamation that goes back decades with the Atlas/Centaur of 1962 and Saturn V of 1967 both relying on combinations of kerosene and oxygen. SpaceX's Raptor came after Merlin and is a family of methane and liquid oxygen engines. Blue Origin has a number of different engine models that use combinations such as kerosene and peroxide; the company went to methane and liquid oxygen for one of its later models, and also has a liquid hydrogen/liquid oxygen engine. Kerosene and methane are fuels and both liquid oxygen and hydrogen are cryogenic propellants, although hydrogen can also act as a fuel.

Ursa Major Technologies founder Joe Laurienti was at SpaceX and Blue Origin before he formed his company in 2015. The company debuted the first of its proposed rocket technologies at the end of that year, but it was another year and a half before they had a qualified engine.[6] So from idea to product, more than three years had elapsed. It's the kind of timeline typical for young space companies such as the ones covered in Chapter 3.

He zeroed in on small launch vehicles and proceeded on the hypothesis that even the large transportation companies would see the economic sense of outsourcing rocket engine production to an expert. The premise flies in the face of how SpaceX and Blue Origin operate: They design it all, earning designations as soup-to-nuts companies. Yet the so-called old-school companies still acquire innovations from startups who can prove their technology is space-qualified.

As with all the other space companies described in the book, there is a market-driven reason why they have a shot at exospheric financial success. Their model meets current performance requirements of space travel: relatively light, cost-effective, and fuel-efficient. Of the three Ursa Major engines, two are flying and one is in development. The two current models are liquid oxygen and kerosene engines, and the one on the drawing board is a liquid hydrogen engine.

In theory, fueling stations could be established in space and supply propellants such as liquid oxygen and hydrogen. Water extracted from asteroids or the Moon would be separated through electrolysis or photoelectrolysis—the latter relying on the Sun to fuel reactions instead of electricity. Spacecraft running low could go to a station and fill up, thereby reducing the importance of tank size in vehicle design. It will happen one day, but of all the projections made in this book about what we are likely to see in the coming decade, fueling stations in the sky may not be one of them. Getting them up and running means maintaining a mining operation that extracts water, and an efficient and cost-effective means of separating the molecules, then storing them at very low temperatures. Hydrogen is liquid at −253 degrees C (−423 degrees F) and oxygen needs a cool −183 degrees C (−297 degrees F).

Space transportation companies realize they must explore other options in the meantime.

GOING NUCLEAR

NASA's nuclear thermal propulsion program is a boomerang. Originally called the Nuclear Engine for Rocket Vehicle Application (NERVA), it was a nuclear thermal rocket (NTR) engine development

program and ran from 1955 to 1973. Considered a huge technological success, it died at the hands of the Nixon administration,
briefly came back to life in 1983 during discussions of the Strategic
Defense Initiative ("Star Wars"), faded again until the early 1990s,
and then came zinging back into the Federal budget after a NASA
Marshall Space Flight Center study showing the value—perhaps
even necessity—of NTR in interplanetary travel. In a May 2019
bill appropriating $22.3 billion for NASA, Congress included
$125 million for nuclear thermal propulsion development.

Although chemical propulsion took Apollo astronauts to the
Moon and could conceivably take humans to Mars, the technology
has huge disadvantages. Chemical rockets are slow and produce relatively little power compared to NTRs—a bit like the family van versus a sports car. According to a 2011 report by a team from the Los
Alamos National Laboratory:

> Nuclear rockets are more fuel efficient and much lighter than chemi-
> cal rockets. As a result, nuclear rockets travel twice as fast as
> chemical-driven spacecraft. Thus, a nuclear rocket could make a trip
> to Mars in as little as four months, and a trip to Saturn in as little as
> three years (as opposed to seven years). Such condensed trip times
> would help reduce astronaut and instrument exposure to harmful ra-
> diation emitted from the cosmic rays and solar winds that permeate
> interplanetary space.[7]

A nuclear thermal rocket uses a nuclear reactor to heat propellant to a high temperature. The propellant is then expanded by a
supersonic nozzle to produce thrust, much in the same manner as a
conventional rocket engine. A low molecular weight propellant
may be used in conjunction with the nuclear fuel. The concern with
NTR, of course, stems from the use of radioactive components. In
Chapter 1, we took a quick look at NTR, suggesting that setting up
an R&D facility on the Moon would remove the sense of danger

many on Earth have about nuclear energy. Blaine Pellicore, Vice President of Defense for Ursa Major Technologies, describes an alternative scenario: "Build the engine on the Earth, launch it into space as a payload, then have the reactor go critical once in space. That would be easier to do than building the whole thing on the Moon."[8]

Whether it's built on the Moon, in LEO, or on Earth, we are striding toward the day when NTRs will aid interplanetary travel. The Mars 2020 rover is using a radioisotope thermoelectric generator (RTG) to power its mission on Mars. NASA also has work underway on a small nuclear fission reactor to deliver power during long duration stays on celestial bodies. Having passed key ground tests already, it may be ready for tests in space by 2022.[9]

REUSABLE LAUNCH VEHICLES

"We have lift off " is NASA's way of telling the world that a space vehicle has ascended from the launch pad. We generally associate this method of departure with expendable launch vehicles (ELVs). They use their entire fuel supply to get into space, but once there it's smooth sailing; whatever is no longer needed for the trip home burns up on reentry or is cast off in space. They have a simpler design than reusable launch vehicles (RLVs), which have to carry a lot more weight than an ELV so they can return to Earth. Rather than discard rocket stages sequentially, RLV leave and come back to Earth with wings, recovery mechanisms, landing gear, thermal protection, propulsion systems, and reinforced systems that can withstand the fatigue of repeated usage.

Once those problems of weight and durability are addressed however, we will probably hear the launch phrase "Take off " rather than lift off, calling to mind an aircraft on a runway that launches and lands horizontally.

Spaceplanes, or Rocketplanes, designed for suborbital space-flight captured the imagination of space aficionados when Scaled Composites won the Ansari X Prize in 2004. At stake was a $10 million prize for the first private sector organization to launch a re-usable piloted spacecraft past the Kármán line twice within two weeks. No government funding was allowed. Scaled Composites, backed by Microsoft co-founder Paul Allen and with Burt Rutan heading both the company and the design effort, developed *Space-ShipOne*, a suborbital spaceplane launched off a subsonic aircraft carrier called *White Knight*. After being released at about 46,000 feet, *SpaceShipOne* soared to the 62-mile (100-kilometer) goal. Vir-gin Group CEO Richard Branson saw the potential in a space rocket that launched horizontally and backed manufacturing of *Space-ShipTwo*, designed for space tourism. Chapter 8 on space tourism picks up Branson's story.

One other aircraft belongs in the company of Rutan's, but it had been retired for thirty-four years by the time Scaled Composites won the X Prize: The North American X-15. It was the first fully reusable space vehicle. The X-15 belonged to a series of experimental aircraft jointly operated by the US Air Force and NASA. Manufactured by now-defunct North American Aviation, the X-15 was a sleek, needle-nose jet, carried up to an altitude of 8.5 miles (13.7 kilometers) by a B-52 and then drop launched. In 1963, NASA pilot Joseph Walker was the first to take the X-15 past the Kármán line; he did the same on his second flight. At the altitude he achieved, Walker could expe-rience weightlessness and see the curvature of Earth. Although not high enough to achieve orbit, the X-15 did reach suborbital space, and was able to land and fly again after some refurbishments.

Ironically, as much as reusability is a primary goal of spacecraft today, it was X-15's reusability that doomed it. The refurbishments, the constant repairs and replacement of parts prompted NASA to cancel the X-15 in favor of expendable rockets. The X-15 was simply too expensive to maintain. Nonetheless, its legacy is not only carrying the first official astronauts into space, but also setting the record for the highest speed every recorded by a piloted, powered aircraft. In October 1967, Air Force Colonel William J. Knight flew it at Mach 6.7—a speed of 4,520 miles per hour (7,272 kilometers/hour).

After the X-15 program was canceled in December 1968, it wasn't until Burt Rutan's success with *SpaceShipOne* that a spacecraft designed for suborbital flight—and not involving an explosive lift off—gave us a sense of next-generation flight.

Rutan is a legend in the aerospace community, earning a reputation for innovations in light, strong, energy-efficient aircraft. He designed Voyager, the plane piloted by his brother Dick Rutan and Jenna Yeager that set an aviation record in 1986 by flying around the world without stopping or refueling. Envisioning *SpaceShipOne*'s 2004 victory as the start of a movement in suborbital flight, he expressed disappointment when, one by one, the other twenty-five competitors for the prize dropped out of the industry. At a 2017 press conference, he told X Prize Foundation chairman Peter Diamandis, "What the hell happened? We got nothing done!"[10]

Although the X Prize competitors did not have Rutan's momentum (translate: financial backing of a billionaire) to carry his innovations forward, other initiatives did take shape after *SpaceShipOne*'s 2004 victory. The impetus for RLV is cheap access to space (CATS), and industry progress like Rutan's indicates points to both single stage to orbit (SSTO) and two stage to orbit (TSTO or 2STO) vehicles, with SpaceX providing a well-known example of the latter..

SINGLE STAGE TO ORBIT

A single stage to orbit vehicle is the process of a ship lifting off from Earth to Earth-orbit in one stage, without throwing away any boosters. One question might be, "Why is it so hard to accomplish this feat?" Many attempts have demonstrated why.

In order to reach Earth-orbit with the best engines available (using the Space Shuttle main engines as a baseline), a launch vehicle's mass must be 85 percent fuel. If the payload is 5 percent of the launch mass, with the remaining 10 percent the vehicle itself, you run into severe limitations on the vehicle design.

In order to reduce the required fuel mass—and consequently, the size of the vehicle—better engines and materials are required. But developmental costs are high, and supplemental government funding would be a key ingredient for these projects, since the cost of R&D is so high.

One major step would be for the demand for launches to be greater than what it has been. If an SSTO launch vehicle can deliver payloads for half the cost of an expendable launch vehicle, then it will be able to capture a huge percentage of the launch market, and generate enough profits to cover the R&D costs of the SSTO vehicle— but only if there were increasing demand for commercial launches. This demand must be high enough for a private launch company to take the risks involved.

There have been developments of high-strength-to-weight materials that could make the SSTO vehicle possible:

- Carbon fiber composites
- Lightweight aluminum-lithium alloys
- Other "exotic" materials, allowing for a strong vehicle structure with a greatly reduced weight. One of those is a ceramic composite, which can wrap around a vehicle to protect it from the assault of re-entry.

The British company Reaction Engines Limited (REL) has a series of SSTO designs called Skylon in various stages of development, with ground-based engine tests scheduled for 2020 and test flights for 2025.[11] One feature of this relatively light design is a reliance on a ceramic composite skin for reentry protection. Initially at least,

Skylons will be unpiloted; they are projected to have the capability of lifting 33,000 pounds/15,000 kilograms to LEO. The program was initially privately funded but has since been subsidized in part by both the British government and the European Space Agency (ESA).

The vehicle is propelled by the Synthetic Air Breathing Engine, or SABRE. The fuel is liquid hydrogen, with liquid oxygen serving as a coolant in Earth's atmosphere, then as a catalyst as the ship heads into space. From a runway to an altitude of 26 kilometers (16 miles), the air breathing engine propels the plane. As the ship soars into the atmosphere, air enters the engines as liquid hydrogen fuels them. At a speed of Mach 5.4, the friction and ramming of the air into the already heated engines will increase the temperatures to over 1,000 degrees Celsius (1,830 degrees Fahrenheit). In order to prevent the engines from melting, liquid oxygen is deposited on the engines, cooling them to −238 degrees Fahrenheit in 0.01 seconds.

As the planes surpasses 26 kilometers in altitude, the air breathing engine shuts down and the rocket engine kicks in, with the liquid hydrogen and oxygen mixing producing a propulsion reaction out of the exhaust, accelerating the plane all the way to LEO.

After its assigned task is performed in LEO, the plane will reenter Earth's atmosphere, like the Space Shuttle, protected by its heat shield. A Skylon spacecraft is expected to have a two day turnaround time, with each individual plane having a maximum use of two hundred flights.

TWO STAGE TO ORBIT

An alternative to the SSTO, and a step toward it, is a two stage to orbit vehicle. The ship consists of two stages, both piloted. When the ship reaches the edge of the atmosphere, the top part of the ship breaks off and proceeds to Earth orbit while the other part of the ship returns to Earth. Both parts of the ships are fully reusable.

There are three types of these two-stage spacecraft, all intended to land horizontally.

- One rocket would be on top of another vertically, to be launched vertically.
- Both rockets would be beside each other, launched vertically.
- One would ride piggyback on top of another, take off horizontally like an airplane, and the spaceplane would break off and fly to Earth orbit while the other plane would land back on Earth.

The TSTO vehicles could be used for a while, paving the way to SSTOs and possibly, the hypersonic plane.

One of the most promising was XCOR Aerospace's Lynx, which relied on using more wing loading than other proposed designs to turn the craft into a sleek spaceplane. Its design would allow it to zoom off a conventional runway, and then when it reached an altitude past the Kármán line, deliver another piloted craft into orbit. The technology was lauded—and still is—but unlike competitors backed by billionaires like Paul Allen and Richard Branson, XCOR suffered from lack of capital and a business plan that fragmented its focus. Ultimately, a nonprofit devoted to aviation education purchased the assets, with a partner company, Sage Cheshire, continuing to focus development of the company's liquid hydrogen rocket.

HYPERSONICS

Anything moving faster than five times the speed of sound, or Mach 5, is called hypersonic. It could be a piloted aircraft, a missile, a drone—anything. We have had hypersonic objects traveling through the atmosphere since 1960, but "sustained, fully controlled flight through the atmosphere at hypersonic speeds, powered by air-breathing engines, has not been attained," according to a January 2020 article in *Air & Space* magazine.[12]

During the mid-1980s, hypersonic transport from Earth to LEO grew popular. This concept holds appeal because hypersonic aircraft would offer quick and easy access to space, with a quick turnaround time and a high specific impulse. They also offer shorter duration flights on Earth. Hypersonic planes would take off from a conventional airport, fly at Mach 25, reach LEO, and land at any other airport, taking off again immediately. The concept is also attractive to those who wish to launch satellites with no waiting period, send up modules to build a space station, or visit a space facility in a shirt-sleeve environment.

In the United States, companies aided by some government funding worked on the X-30, a two-man hypersonic vehicle for the military, which in turn would have led to the development of the National Aerospace Plane (NASP). As of the late 1980s, the United States, Germany, Japan, Britain, France, and Russia made proposals for their own hypersonic vehicles. Unfortunately, the US hypersonic program was canceled in 1993 due to a lack of perceived necessity and cuts in the federal budget. Prior to cancelation, extensive computer calculations were conducted showing how a hypersonic system could work. It came down to three engines: turbojet, ramjet, and scramjet. The turbojet launches the plane off the ground to speeds of Mach 2.5. The ramjet then takes over and flies from Mach 2.5 to Mach 5.5. The scramjet then comes in and flies at speeds up to Mach 25. This is how each type of engine works, and why they are all needed for hypersonic flight:

- **Turbojet engine.** Used in most subsonic and supersonic aircraft, it sucks air in the form of oxygen and nitrogen into the inlet and compresses it. Fuel, consisting of hydrocarbons, is injected into the combustion chamber and burned. The expanding hot gases surge out of the nozzle producing thrust. The gases spin the turbine, which powers the compressor. It maxes out at roughly three times the speed of sound, or Mach 3, because it relies on subsonic air.

- **Ramjet engine.** An engine that cannot move an object from standstill, but needs forward motion to kick into action. Ramjets display their power between Mach 3 and Mach 6. The normal ramjet inlet slows down the incoming air while compressing it, then burns the fuel and air subsonically and exhausts the combustion products through the nozzle, thrusting the plane to its higher velocity.

- **Scramjet engine.** The name stands for supersonic-combustion ramjet engine. It ingests air through a specially designed inlet and mixes this with hydrogen, carried aboard as fuel in liquid form. The hydrogen and oxygen, taken from the atmosphere, are ignited and thrusts the engine at hypersonic speeds. The exhaust product is water. "In a scramjet, the air comes into the engine and you don't compress it . . . you design the inlet so it keeps moving at supersonic speed."[13] A key result is that temperatures don't rise as high as with a ramjet engine, so the process of injecting fuel, mixing it, and burning to get thrust can be sustained.

Hypersonic flight is vital to the future of space transportation, and tests with prototypes are ongoing at places like the US Air Force Arnold Engineering Development Complex, which is home to a hypersonic wind tunnel. In the meantime, Boeing has its eyes on development of a Mach 5 airliner to fly in the *transatmosphere*, a region of the atmosphere between about 120,000 feet and 500,000 feet, "a border area between flight within the atmosphere and flight into space."[14]

GROUND TRANSPORTATION

In *Star Wars*, the giant robots with articulated limbs were known as All-Terrain Armored Transports (AT-ATs) and had combat functions on behalf of the bad guys. In real life, they are called ATHLETES and they are designed to grip, dig, carry passengers, and otherwise serve good-guy functions on celestial bodies.

ATHLETE stands for All-Terrain Hex-Limbed Extra-Terrestrial Explorer, and it's a product of the Jet Propulsion Laboratory's (JPL) ingenuity. In describing its ATHLETE Rover, JPL notes:

> ATHLETE uses its wheels for efficient driving over stable, gently rolling terrain, but each limb can also be used as a general purpose leg. In the latter case, wheels can be locked and used as feet to walk out of excessively soft, obstacle laden, steep, or otherwise extreme terrain. ATHLETE is envisioned as a heavy-lift utility vehicle to support human exploration of the Lunar surface, useful for unloading bulky cargo from stationary landers and transporting it long distances.[15]

Mass drivers could eventually complement a vehicle like ATHLETE. Proposed by the late Gerard O'Neill, the Princeton physics professor most associated with habitat designs (see Chapter 7), a mass driver would support large-scale transport of lunar material by launching small payloads.

O'Neill envisioned it as an electrical device that runs for several kilometers on a track along the lunar surface. Payloads of any material are accelerated to a high velocity along the track. Small vehicles, called buckets, contain superconducting coils to carry these payloads. These buckets are accelerated by pulsed magnetic fields and are guided by inducted magnetic fields set up in a surrounding guideway. Upon reaching the correct velocity, the buckets release their payloads and accelerate into space toward Earth. The buckets are then slowed for recirculation and reuse.

At a fixed point in space, (possibly at L2), there is a mass catcher, to catch the material ejected by the mass driver. The material is then retrieved and sent to either LEO for processing, or to Earth. A mass driver would save transportation costs and propellant compared with rocketing material to Earth.

Many designs for rovers and "taxis" have surfaced through the years, but one unseen part of all of them is the brain that keeps them responsive to human beings—software. Compensating for signal latency that would affect a remote operator trying to control a vehicle on the Moon or Mars surface is just one example of the challenge. Another is providing a 3D view of the environment, whether the operator is remote or in the lunar vehicle.

Olis Robotics develops software that operates robotic systems remotely, and it has won some prime subcontracting responsibilities from NASA and other government agencies due to the team's experience with robotics in extreme environments. Before venturing into the harshness of space, Olis Robotics demonstrated expertise in deep water, which presents latency challenges like space. Signal delays can sabotage a mission, with robots failing to "hear" instructions at critical moments; Olis Robotics has focused on streamlining the link by solving for latency. Among other applications, Olis's software has an integral plug-and-play role as part of robotic arm on the Maxar

Technology satellite designed to collect rock samples for NASA on the Moon.[16]

Formerly called BlueHaptics, Olis Robotics spun out of work at the electrical engineering and applied physics lab at the University of Washington. The team that evolved reflects expertise in robotic surgery, machine learning, and complementary disciplines; when applied successfully to deepwater missions, NASA determined they had also earned their credentials to work in space.

The company won the SpaceCom $100,000 Entrepreneur Challenge in 2019. CEO Don Pickering presented the winning case for the company's innovations by explaining their work in providing a realtime, navigable model of the environment; 3D model generation using a single secure digital camera; and motion compensation capability, among other technological advances. Pickering explains:

> The legacy of robotics has really been one that's mechanical. But there are challenges that can only really be overcome through software in terms of getting robots to do precise work. To do that, we need to use sensors and algorithms and software that can allow remote robots to use realtime 3D information—to have the ability to recognize objects and then have humans calibrate modifications so the robots do sophisticated work.[17]

THE NEED FOR NONTECHNOLOGICAL INITIATIVES

During the presidential administration of Barack Obama, the United States took a dramatic turn toward relying on private companies to take us from Earth to heavenly bodies. Never again would Americans think of NASA as the sole source of space transportation.

In 2010, President Barack Obama made the decision to put future US space transportation to Low Earth Orbit, especially to the International Space Station, in private hands. This may turn out to be one of the best decisions made for the space program since Apollo. The space community applauded the move for lowering the costs of a flight—that is, making the investment in space transportation a proposition that makes sense on a spreadsheet. From the perspective of Casey Drier, director of space policy at the not-for-profit Planetary Society, the move to embrace the potential of commercial launch capabilities is "the biggest legacy"[18] of the Obama administration.

Luxembourg and the United Arab Emirates implemented similar initiatives supporting commercial transportation efforts in space.

The international community needs to follow suit, recognizing that agreements such as the Outer Space Treaty set the stage for cooperation and need not be abandoned, yet the economic exigencies of—and bold, respectable visions for—space transportation put private enterprise in a central and critical role.

SIX

GOING TO WORK IN
NEAR-EARTH SPACE

Working in space can mean being physically located in space or on a celestial body while doing a job, or it can mean working remotely from Earth. For example, researchers at BioServe Space Technologies, a research center within the University of Colorado, can monitor experiments on the ISS by seeing the live feed from their microscope in a control room in Boulder, Colorado.[1] Another example of remotely working in space is directing robotic repairs or manufacturing from a facility on Earth.

The first part of this chapter looks at Earth-focused activities that could not occur without the ability to go to space. They involve research and manufacturing intended primarily to improve life on Earth. After that, we look at the Earth-based activities needed to enable, complement, and support space exploration and settlements on celestial bodies.

Then we launch, exploring current and near-future jobs performed in off-Earth environments. We try to keep it real by staying within this solar system and relatively close to home—Low Earth Orbit and the Lunar surface—and within a time frame of roughly 2020–45.

DOING SPACE WORK ON EARTH

Earlier in the book, we introduced the kind of research and manufacturing going on in space now. Taking a closer look at R&D projects at specific companies, universities, and space agencies yields more specifics on the breadth of career opportunities in the space industries now and in the near future. The companies range in size from about ten people to thousands, and the R&D spans disciplines such as horticulture, medicine, communications components, small parts manufacturing, and much more. The universities are public and private, with California Institute of Technology, Harvard University, University of Cambridge, and the Sorbonne among the top places to do space-related research. Other universities are top-ranked for specialties such as space medicine and space mining. There are about eighty countries with space agencies as of this writing, although their capabilities vary significantly. Some have astronauts; some do not. Some have launch capability; the great majority do not. Most, but not all, operate satellites.

SMALL COMPANY, BIG VISION

Space Tango is a Kentucky-based company with investigations touching on multiple disciplines. Twyman Clements is the CEO of the company, mentioned in Chapter 1 as being behind innovative activities aboard ISS.

As a junior in mechanical engineering at the University of Kentucky, Clements began working at a CubeSat lab.[2] CubeSats are small cubes containing research projects done in space. They date back to 1999 and grew out of a NASA-supported collaborative effort between faculty members at California Polytechnic State University (Cal Poly) and Stanford University's Space Systems Development Laboratory. The original intent was to help universities develop space programs by lowering the barriers for using the ISS for complex science and manufacturing. Because CubeSats in their most basic form are made of low-cost parts, even middle schools eventually skipped into space-based research.

After graduate school, Clements managed the CubeSat program at the same lab. From there, he went commercial in 2014, partnering with Kris Kimel, then CEO of Kentucky Science and Technology Corporation. They formed Space Tango, which has taken the CubeSat concept to a significantly higher level, combining multiple compartments for different experiments in a CubeLab. A combination of automation, fluid flow, temperature control and other factors allow for many different types of processes to occur within a single CubeLab. The cubes vary in size: The standard is 1U ($10 cm^3$), and from there, they scale upward to 2U, 4U, 6U, and 9U.

The diversity of Space Tango's collaborative projects illuminates how many different types of professionals are contributing to the R&D. It also suggests a plethora of new jobs that will emerge when the work progresses from the research stage to manufacturing new products in microgravity.

The experiment Space Tango conducted that was referenced in Chapter 1 involved growing barley seeds in zero gravity for Anheuser-Busch. Although we jokingly suggested that early space settlers could look forward to craft beer, the company's follow-up to the original experiment suggests that's not a joke. On the next round in late 2019, the Space Tango team took the dried barley and malted it, which is a more complex process involving various temperature changes, submerging it in water, and more. It was a big step closer to "Bud on Mars" and a career as a lunar brewmaster.

The experiments support work of Anheuser-Busch's barley research group by analyzing the epigenetic changes to barley in zero gravity. There is a strong marketing relevance to the effort, though: It promotes brand awareness. The company has said, "When we get to Mars, Budweiser will be there" and the collaboration with Space Tango gives credence to the claim.[3]

Beer on Mars aside, Space Tango's mission is to develop and produce products in microgravity and return them for use on this planet. Clements says,

> What Space Tango produces is not intended to contribute to space infrastructure; it's focused on Earthly needs—using space to improve infrastructure here on Earth. Everything Space Tango works on is made in zero gravity for Earth. We're not building things to expand into the solar system.[4]

The biotechnology experiments and early manufacturing efforts exemplify this aim. Jumping onboard the SpaceX Commercial Resupply Services 16 (CRS-16) in December 2018, Space Tango brought an experiment created with LambdaVision to evaluate protein-based retinal implant production. The objective was improving the quality of protein-based retinal implants to help patients blinded by retinal degenerative diseases, such as retinitis pigmentosa and age-related macular degeneration.

Clements sees R&D in microgravity as a value-adding proposition, and his description of the benefits of the retinal implant work express that view in specific terms:

The improved microgravity manufacturing paradigms have the potential to increase the stability, activity, and optical quality of the prostheses, reduce the amount of raw materials needed for assembly, and accelerate production of the devices for further preclinical and clinical testing.[5]

The same technological advantages to retinal implant development that Clements describes are applicable to other products. Space Tango's collaborations on stem cell tissue engineering and tissue chips illustrate the way microgravity enhances biological R&D processes. A key example is that microgravity can effectively maintain *stem cell pluripotency*, meaning the ability of the cells to differentiate into other specialized cell types. The more cell types they can differentiate into, the greater their potency. And the tissue chip research—the chips are artificially grown bits that mimic the function of an organ—has potentially enormous implications for disease control. These chips emulate the complex biology and mechanisms of an organ outside of an animal for drug-screening and other applications. Space Tango has collaborated with companies on lung chips, blood-brain barrier chips, and even brain organoids. Like tissue chips, Clements describes brain organoids as "analogs for actual brains—miniature organs, artificially grown yet resembling the brain. They don't have consciousness or complex thought, but neurologically, they do what the brain does."[6]

As Space Tango has gotten more customers and more complex work, experience has supported Clements's assertion that microgravity is a value-add. The company has used the ISS and SpaceX equipment to build a portfolio of diverse CubeLab investigations to determine what the technologies need to be to go from raw material to finished product. He also notes: "If the Space Station is decommissioned or otherwise goes away, we have an idea to move the activities to another platform—a factory—to build those products at scale."[7]

A UNIVERSITY PARTNER IN CURING DISEASES

Dr. Luis Zea is an assistant research professor at BioServe Space Technologies, a research institute within the University of Colorado in Boulder (CU). BioServe enables scientists' experiments to be performed on the ISS by developing the necessary hardware, modifying the protocols so they can be done in space, and training the astronauts on board the ISS to run the experiments. In certain situations, Zea and others on his team also do the ground controls from their lab in Boulder.

In his view, there is a twofold approach to microgravity research in the field of bioastronautics. One is to enable the safe exploration of space by humans. Myriad factors that we know of can be detrimental to human space exploration, such as bone and muscle reabsorption and aberrant behavior of certain types of bacteria. The other aspect, which is primarily what Zea focuses his research on, is using the microgravity environment of space to find solutions to medical problems on Earth. He explains:

> When you think about it, life has evolved for at least three and a half billion years on our planet where temperatures have changed and the composition of the gas and salinity of the oceans—all these things have changed in this period of time but gravity has been a constant. So when you have the opportunity to bring a cell or an organism to an environment like the microgravity of space, now you can see processes that you cannot see on Earth. That unmasks molecular processes that, when we try to replicate them on earth, we don't see them again.[8]

Zea's approach treats gravity as an independent variable. He tests hypotheses in microgravity with the ultimate objective of finding new ways of solving perennial problems such as drug resistance of organisms.

Two experiments running concurrently on the ISS particularly piqued his interest. Both were for startups, one in Texas called n3D Biosciences and one for a startup in Massachusetts called Oncolinx. For the Oncolinx experiment, the ground controls were in a room at BioServe. He watched the astronauts performing the experiment in realtime, and when they had completed the tasks, he ran down to his lab to replicate what they had done with the ground controls.

The cells used in this experiment were from a human adenocarcinoma from a twenty-six-year-old male. Adenocarcinoma is a type of cancer found in glands, but it can spread to other parts of the body. What Zea observed was eye-opening. In a microgravity environment, researchers were able to see structures clearly, such as tumors forming. On Earth, they didn't get the true picture because gravity flattens the structure, whereas the structures are 3D when gravity isn't pulling them downward.

When you put the cancer cells under a microscope here on Earth, you change the focus on the microscope until you have the cells in focus. As soon as you move the focus, say, a little bit downward, you lose focus of everything. You bring it back up, you see everything is in focus. You move it slightly up, and again, everything is out of focus. Everything is bi-dimensional. As soon as another cells goes and puts itself on top, gravity will bring it down and turn the cells into something like pancakes.

But when we see the astronaut doing that experiment—and we were seeing the live feed from our microscope—we could see how it traveled on all three axes. We could continue seeing the gigantic structure of these adenocarcinoma cells building a 3D structure.

The Oncolinx experiment was testing a drug called Azonafide. The novel aspect of what they were doing was combining it with an antibody that would behave like a puzzle piece, where on one side it would bind only to the drug and on the other side, it would bind only to the unhealthy cell. That's why they wanted these three-dimensional structures that you can only see in microgravity.[9]

Why can't we do this on Earth? Attempts to replicate this on Earth might be done with a random positioning machine or a clinostat, but they necessarily involve frustrations. The vessel rotates on one axis, the whole thing repeats on a second axis, and then the whole system rotates on the third axis, so the researcher is continuously rotating the gravity vector. The nature of the process limits how big it is possible to grow something.

Another of Zea's keen interests is inhibiting osteoporosis. This is a condition affecting astronauts profoundly, and a drug or process able to help them could translate into enormous benefits for middle-aged and older people on Earth. The worst case of bone reabsorption on Earth is about 1.2 percent of bone mass a year for post-menopausal women if it goes untreated. In space, if astronauts were not to take any drugs or work out one or two hours a day, they would see about the same amount of bone absorption *a month*. That's a problem for space exploration, but also an opportunity to test molecules to see if they can go through the pipeline and eventually become drugs against osteoporosis either by inhibiting bone reabsorption or promoting bone generation. Experiments in the design stage don't involve experimenting on astronauts, but rather on mice in space. Some will be treated with a molecule designed to abate osteoporosis; some wouldn't get treated. At the same time, a control version of the experiment will be replicated on the ground.

In a university research facility like BioServe, where all the partners are entrenched in some aspect of human physiology yet have diversity in their prime objectives, professionals and graduate students from multiple disciplines find a place to thrive. For example, some of the research at the CU facility work on human-machine interfaces. To do that, they need people savvy in computer code. For other projects, a background in environment controls and ecosystems is required. A physical-chemical approach to a problem, such as examining chemical reactions to extract oxygen or to remove CO_2, requires people with a chemistry background. For Zea's primary area of work, space microbiology, he uses students with backgrounds in microbiology and aerospace engineering, although most of the work is in the microbiology lab. As he suggests in his advice to students looking to develop a career in bioastronautics:

> Bioastronautics is a multi-disciplinary field. In space microbiology, you need to know microbiology as well as some aspects of aerospace engineering. In space biomining, you need geology and microbiology. If you want to do "human factors" work, which involves the design of spacecraft and habitats, you need some background in psychology because you want to make sure the machines you're building are operable—that the people using them are happy using them. You don't want them hating an environment or piece of equipment that they will be spending a lot of time with.[10]

We pick up the discussion of biomining in Chapter 9. In brief, it's getting bacteria to eat dust and turn it into a precious metal.

CANADA'S LONG ARM

Canadarm, also known as the Shuttle Remote Manipulator System, is considered Canada's most renowned robotic achievement. It served shuttle missions in space for thirty years, before being retired when the United States ended the shuttle program.

> Canadarm was a really fantastic piece of hardware. We didn't really have the whole software picture put together (in the early 1980s). We knew we had this really great widget that we kind of threw over the fence and said, "Hey, NASA! You should use this on your space shuttle."[11]

This is how Ken Podwalski, director of Gateway for Canadian Space Agency describes CSA's entry into a space partnership with the United States. For a while, Canada was treated like a junior partner in that NASA insisted on handling the remote operation of the robotic arm. Ultimately that changed, with Canada assuming operational control, causing Podwalski to conclude: "They (NASA) elevated us and they helped us."[12]

Canadarm 2 succeeded the original on the ISS. A 17-meter-long (56-foot) robotic arm, it was built to be reburbished in orbit and, in fact, had one of its wrist roll joints replaced in 2002 by a spacewalking astronaut. Canadarm 2 had a key role in the assembly of the orbiting laboratory, helped with station maintenance, moved supplies and equipment, and performed "cosmic catches" by grappling visiting vehicles and berthing them to the space station.

Canadarm2, the International Space Station's robotic arm, as seen by CSA astronaut David Saint-Jacques at sunset. (Photo courtesy of the Canadian Space Agency; credit CSA/NASA) © Canadian Space Agency.

Canada reinvigorated its partnership with the United States in 2019 by joining NASA's Lunar Gateway Station Project with

Candarm 3, which will have a critical role in construction. In fact, Canada has the lead role in the robotics related to Gateway development. As Prime Minister Justin Trudeau tweeted (with emojis of a rocket and a moon added):

> Canada is going to the moon! The new Canadarm 3 will assemble & maintain the Lunar Gateway, & it will be made in Canada, by Canadians—keeping us at the forefront of innovation & creating good, well-paying jobs.[13]

Canadarm 3 will have a small sidekick to carry out robotics tasks requiring dexterity, a model similar to Canada's Dextre robotic "handyman" on the International Space Station. And while crews assigned to Gateway are working offsite, Canadarm 3 will continue handling day-to-day tasks on its own. That level of autonomous operation is precisely what will be required of robots assigned to construction, repair, mining, and countless other heavy-lifting chores to support a human presence on the Moon and Mars.

WORKING WITH ASTEROIDS

Apollo missions were just about winning a race. By the early 2000s, that mindset about venturing into space had changed, with visionaries planning ways for humans to remain in space and on celestial bodies.

At that point, the Colorado School of Mines (Mines) started doing research on how to collect, excavate, and extract resources, mostly for oxygen production. More than a decade later, a couple of companies announced plans to get into asteroid mining, and years later the interest on extracting and utilizing all sorts of space resources extended into more companies and internationally. And then the decision was made to develop programs to train professionals who would work in that field. In 2018, Mines launched the first program in the world focused on educating scientists, engineers, economists, entrepreneurs, and policymakers in the developing field of space resources.

The topic of asteroid mining caught the attention of media, which started talking about mining asteroids for metals as though it were a recreation of the Gold Rush of the mid-nineteenth century.[14] Since then, others have entered the field with an interest in doing this, not necessarily to bring resources to Earth, but to explore in-situ-resource utilization (ISRU). They aim to use resources found in space for space applications.

As we noted earlier in the book, asteroids offer enormous mineral wealth. However, in terms of practical next steps, gathering rocks of great value on Earth is taking a back seat to harvesting water. Extracting water from bodies like asteroids and the Moon are part of important ISRU efforts, along with harnessing resources at the exploration site to build habitats and other infrastructure required for extended stays in space, and ultimately settlements.

Dr. Angel Abbud-Madrid, director of the Center for Space Resources at Colorado School of Mines, says their work centers on using knowledge to help crews stay longer on planetary bodies.

> Our focus is to help them use whatever is out there to live off the land. Bringing things to Earth is not the priority right now. It's about allowing space travelers to not carry everything from Earth, which is very costly, but it also demands a lot of energy to send materials from Earth to space.[15]

Interest in resource mining has escalated internationally since about 2010, both for space agencies and companies. Although NASA was a prominent player in the beginning, space agencies in Russia, China, Japan, South Korea, and Europe have activity going in this area. Luxembourg is also positioning itself as a favorable corporate base for space business, much in the way that the state of Delaware positions itself in the United States.[16]

The business case for asteroid mining companies has always been a long-term proposition. Planetary Resources—calling itself "The Asteroid Mining Company"—got a lot of attention initially for attracting investment money to its bold vision of extracting resources from asteroids, principally harvesting water for use in space. It was acquired by ConsenSys in October 2018 and the focus changed more generally to "space initiatives," according to their website.

PHASES OF DEVELOPMENT

Dr. Abbud-Madrid helped us make some projections about what the job market in space activities might look like over the next five, ten, fifteen, and twenty years, with a focus on asteroids and Lunar resources. The development of opportunities tracks with the milestones that will be achieved:

- Identifying the resources
- Developing the technology to extract them
- Large-scale production and full commercialization of opportunities

From about 2020 to 2025, most of the jobs related to asteroid mining and space resources in general will be here on Earth and involve the prospecting phase of the process—that is, identifying the nature, location, and quantity of resources. Remote sensing satellites would circle the asteroids and Moon and figure out how much water is there, for example. The data sent back to Earth will answer questions such as "How deep is it?" and "How clean is it?" Rovers traversing the surface will probably also be part of the prospecting process, with equipment on the surface involving excavators and drilling systems.

It's important to note that we have to get better at using what is already in space at the same time we're building equipment with new applications. Satellites already in space, and the ISS, as long as it exists, have a role in preparing to mine celestial bodies and expand our exploration well beyond the Moon. And as much as it has a terrible reputation, one of the most important resources in Low Earth Orbit may be space debris. The material we've sent into space for decades that is still there should turn out to be a resource. Of course, using it is contingent on reliable manufacturing processes to take advantage of a high-vacuum or ultra-high-vacuum environment with low gravity or no gravity. Success in asteroid and lunar mining necessitates that expertise: It's a straight path from extracting resources to making products that are usable on-site, as well as processing resources intended for use on Earth.

One of the first companies making inroads in space manufacturing is aptly named Made in Space. They were able to put the first 3D printer in space (as described in Chapter 1). Now they are looking to make structures in space so we will not have to send heavy equipment and parts from Earth. Combine output from a 3D printer with parts from a dead satellite and a few hunks of space debris and who knows what kind of robotic life might take shape!

With leadership, coordination and funding from space agencies, and some private investment money, early efforts to mine asteroids will most likely engender the development of many startups. They will likely be the ones designing and building those excavators and drills for rovers that will be loaded on to the rockets made by SpaceX, Blue Origin, and others. From engineers designing the equipment to welders putting it together to remote sensing scientists, multidisciplinary work forces will be required to execute what business plans will likely call "Phase I Asteroid Mining: Prospecting."

Phase II will overlap. Let's speculate it will go from 2023 to 2030, as soon as the initial information from sensors generates interest in going after the resources. Investors, whether that means sources of money from the public or the private sector, need reasons to think the identified resources can be transformed from something potentially valuable into something actually valuable.

At least one event suggests the timeline may be shortened. A Japanese mission, *Hayabusa2* spacecraft, made contact with the asteroid Ryugu on July 11, 2019. Reports about the mission to the asteroid Ryugu said that *Hayabusa2* "touched down" on the surface.[17] Not exactly. On the first mission, the Japanese vessel swooped down, fired a bullet into the asteroid surface to loosen material, and then collected samples with a horn-shaped instrument. On the second mission, it fired a larger impactor into Ryugu to create a crater, collected material it had stirred up, came back and fired again, and then snagged a new load of samples.

No follow-up mission was planned, so this was not the start of a Hayabusa series aimed at mining asteroids. The purpose was purely a scientific effort of the Japan Aerospace Exploration Agency (JAXA) designed to collect samples and return them to Earth for analysis. Nonetheless, this kind of research and exploration bridges prospecting efforts with mining projects.

During the period of project development we're calling Phase II, we should see a few companies taking steps toward celestial bodies. Expect companies to launch demonstration projects—proof-of-concept mining activities that extract, store, and process water.

As with the development of any other major undertaking involving the safety of human beings such as building ships and bridges, governments will have a big-picture involvement even though companies may be doing all the work. Governments tend to serve as both an information and a coordination resource to enhance the likelihood of success and minimize the chance that people are put in harm's way in the project, among other things. In terms of the practical details of asteroid and lunar mining, however, governments will probably not have much involvement.

By the mid-2030s, much larger scale production should begin to take shape. The water mined on an asteroid will be processed so that transportation companies have ready access to propellants in space. These are the gas stations of 2040. This is where we can begin setting up a true space infrastructure with way stations and refueling stations to help support space transportation. Another benefit will be refueling satellites so they can keep chugging along in orbit instead of becoming space debris.

The propellant can also be sent back to Low-Earth Orbit or other important orbits between Earth and the Moon and sold to companies to fuel their rockets. At that point, commercialization of space is blossoming. Products that are made in space may be sold in space; some products that are made in space are also "exported" to Earth and sold.

Moving from the prospecting phase to relatively stable commercial operations will involve the usual panoply of administrative and scientific/technical personnel, but expertise will be especially sought after in the areas of advanced robotics and machine learning (artificial intelligence). The scenario we opened with in Chapter 1 with "Mac" managing a workforce of robots on an asteroid would be a human managing a workforce of robots who mine and process resources, handle repairs on other robots, and probably even act as first-responders when a human gets injured. We explore these AI capabilities of robots much more in Chapter 10.

Students and professionals in the scientific and technical fields just covered in this discussion should take note: Other industries want you as well. Abbud-Madrid is uniquely situated at Colorado School of Mines to see what kinds of jobs his space resources students might find in companies that have existed for a century or more:

> The mining and the oil and gas industries, as well as equipment manufacturers, are interested in space mining because, at first, it involves developing technologies that will also be useful to them on this planet. Right now, these companies have a dual purpose: Develop technology that can be used on Earth for current operations—that feeds the revenue to keep the company going—and then eventually, move operations to space.

They are looking for very advanced robotic systems that can go to remote places on Earth, involving substances like highly radioactive material found deep on Earth—very risky for humans, but something that autonomous robots can navigate and process. This is the same technology that will give them access to resources on asteroids and the Moon after being adapted to different gravity and temperatures.

They want the technology here, but they also want to be ready for what comes next.[18]

Additive manufacturing, also known as 3D printing, is also an essential discipline in the commercialization of Lunar and asteroid mining. The ability to do additive manufacturing in microgravity is central to making money in space. To strip down an explanation of additive manufacturing: It involves adding layers of material, in a precise geometric shape, to produce an object. Relying on digital processes versus analog, this manufacturing technique has been used successfully on Earth for decades, but with research such as that done by Made in Space on the ISS, it has proven to translate very well to a microgravity environment. The logical application is to use the regolith on the surface of a celestial body to build tools, spare parts, and even habitats.

Again, companies are interested not only in harnessing such technology for activities in their business plan twenty years out, but also for current use—for profit-making ventures on Earth. So, for example, the same kind of agenda that applies to using additive manufacturing to build in space applies to next-generation power generation and enhanced communications. Abbud-Madrid advises:

> This is where the opportunities arise now. Provide the infrastructure in space that you need to conduct these operations, but do it a way that can go from terrestrial applications to space.[19]

CONTACT WITH ASTEROIDS

Space mining operations will rely on both the technology behind physically interacting with asteroids as well as stabilizing the position of asteroids. *Hayabusa2*'s successful contact with Ryugu literally just scratched the surface of needs to happen to begin prospecting asteroids.

This image illustrates an initial contact with an asteroid to enable sustained interaction with the body. The tethered connection is an attempt to rein the asteroid in so it could be boarded or towed to a more desirable location. A premise of this kind of connection is that an asteroid has a very small gravitational force. The simple scenario of landing, deploying an excavator, and drilling would result in sending bits of rock into space. So how do you extract resources from an asteroid you've just lassoed?

Extracting water from a carbonaceous body, as some asteroids are, bears no resemblance to customary approaches to extraction of water on Earth—although the process can be done on Earth. There is no drilling to find a pocket of water, for example. The water is part of the rock and process of extraction involves intense light and heat.

The process is optical mining, that is, drilling with light. A reflector on a spacecraft would be put at an angle to reflect and focus the Sun's rays. The process applies the same principle learned in wilderness training of starting a fire with solar ignition. In the back country, you would wait until the Sun is directly overhead, then angle a five-power or greater magnifying glass to focus the Sun's beam. The caveat is, don't use a white material as tinder because it will reflect the light and heat.

With an asteroid, a powerful reflector would direct that solar energy at the location where mining is desired. The Sun heats the rock; the rock releases water. But when the hot water boils and begins evaporating, it forcefully tries to escape—the same phenomenon seen when boiling water pushes a pot lid upward. This process of expansion breaks the rock apart.

When the volatile elements, such as water and carbon, bleed out, they are frozen in a process called cold-trapping. Abbud-Madrid explains:

> You end up with a bag covered with ice. You remove the bag and you can refine the water and use it there, or you can bring it to an off-Earth facility where you turn it into hydrogen and oxygen to create a propellant or a variety of other things.[20]

A description of how the process is simulated in the lab clarifies the forces at work. It also highlights the fact that optical mining on a celestial body is a well-understood process, not just a theory on paper.

- In the lab, the extraction involves a high intensity lamp generating a concentrated beam in lieu of sunlight.
- Another element of the system that simulates the environment of space is a vacuum chamber. The same chamber can be adjusted to simulate conditions on Mars, for example; after the air is extracted a little carbon dioxide (CO_2) is pumped into the chamber.
- A third component in the system regulates temperature. Liquid nitrogen is injected into a chamber to bring it down to –190 degrees Celsius (–374 degrees Fahrenheit) to simulate night conditions on the Moon, for example; hot gas is injected to simulate daytime conditions, which would be more like 100 degrees Celsius (212 degrees Fahrenheit).
- To extract water from a mock asteroid or lunar sample, therefore, the rock is placed into the appropriate environment in terms of pressure and temperature and then blasted with the intense light. The water comes out as a vapor and then condenses into ice as it hits freezing temperatures. This step consists of exposing the gas to the extreme cold generated by liquid-nitrogen infused coils and turning the water extracted from the rock into ice. After that, it can be thawed, purified, and consumed, used for shielding from radiation, or separated into oxygen and hydrogen, both of which have value as propellants.

Dr. Angel Abbud-Madrid, Director of the Space Resources Center, Colorado School of Mines, holding water extracted from a simulated asteroid rock through the process of optical mining.

Using a tether to move the asteroid to a more desirable location for mining and processing might mean towing it to a point where the body stabilizes in space. In theory, the way to do this is by using La-Grange libration points, explained briefly in Chapter 1 as positions in our sky having specific properties related to the Moon's oscillation patterns. Instead of mining asteroids in place, they might be moved via solar sail, space tug, or rockets to the Moon's orbital path, 60 degrees from the Moon either way, at points known as L4 and L5.

The five LaGrange Libration Points are where the gravity fields of Earth and the Moon cancel each other out, producing a stable orbit. An object placed at these orbits would feel equal and opposite attractions from both Earth's and Moon's gravities, and stay fixed at that orbit. L4 and L5 are the most stable of these orbits.

Think of the concept this way: An asteroid at a LaGrange Point would be like a marble on a mountaintop—steep slopes on all sides. There is one place at the top of the mountain where the marble has a stable resting place.

Around 1774, the French-Italian mathematician Joseph Louis La-Grange (1736–1813) used Newton's gravitational theory to explore properties of two unique points in Jupiter's orbit around the Sun (L4 and L5, 60 degrees from Jupiter, both behind and ahead of the gas giant respectively). Years later, several asteroids were discovered to be trapped near the LaGrange points in permanent orbit, and these became known as "Trojan" asteroids.

In the Earth-Moon system, the stability of the L4 and L5 orbits are a lot more complex. The Sun, distant as it is, greatly affects orbits in the vicinity of Earth, because of its enormous mass. This disturbs the orbit of any object at these two points.

Fortunately, there is a work-around. In 1969, A. A. Kamel of Stanford University calculated that the stable regions at L4 and L5 lie in orbit of these points, 90,000 miles around the central libration points (see Figure 6.1).[21] The orbit, which is around Earth, would be in an eccentric motion.

The other libration points are L1 and L2, at opposite ends of the Moon, and L3, 180 degrees from the Moon. It is recommended that we tow these asteroids to Lunar, or even Earth L4 or L5 libration points.

The laws of gravity that apply to the Earth-Moon libration points also apply to the Earth-Sun libration points. Sixty degrees on either side of its orbital path around the Sun also lie what might be called the Earth-Sun L4 and L5. Larger asteroids could be deposited at these points for future use.

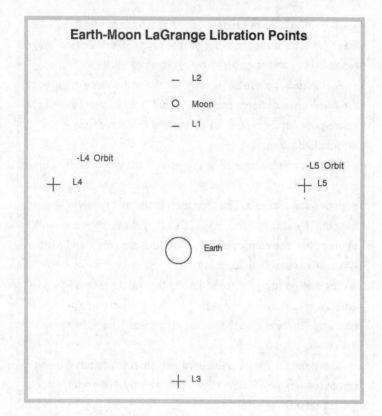

Figure 6.1: LaGrange Libration Points

HANGING OUT NEAR ASTEROIDS

Two nineteen-year-old entrepreneurs steer their spacecraft, owned by their father, into a near-Earth asteroid. They are towing an old, spent rocket.

Jake and Isaac Simms found a livelihood looking for space junk, otherwise defined as heavy debris, and either loading it into the ship's compartment or towing it if it was too big—like old rocket stages, usually found in graveyard orbit, about 23,000 miles above Earth. These stages and dead satellites were getting rare because a lot of scavengers like Jake and Isaac were cleaning them out of that orbit for extra cash. They realized they had to go elsewhere, closer to Earth.

Fortunately, no two ships went for the same piece of junk at the same time, or fights could break out. There was still a lot of space junk out there, but it had started to decrease, and their income had taken a hit.

As Jake and Isaac dock at the asteroid, a line is shot at the stage and the brothers release their ship's line as the stage is pulled to the asteroid. The ship then lands and is towed into a hanger. The area is filled with air as they disembark and walk to the office. The compartment of the ship is opened, and a broken, useless satellite taken out.

The two go to the office to talk to the head honcho, Roy Soames. A breakdown of the satellite has revealed some gold and platinum. There is also the rocket stage, worth something from the scrap.

Analyzing the metal his subordinates have registered on his computer, Roy says, "I'll give you both twenty-five grand."

"Is that all?" shouts Jake.

"Listen, guys, we have to do a lot of work separating these metals before melting. You're doing the easy part by catching them. We're the ones working our butts off!"

"And you're going to get ten times the amount you're giving us."

"That's right! Be glad you're getting what you're getting. Most junkyards won't even give you half of what I'm giving you!"

Isaac sighs: "I guess we have no choice."

"Gonna go out for some more?"

"Yeah, but the graveyard orbit is almost cleaned out. We're going closer to Earth. We can at least get some junk satellites."

"They're harder to come by, even in Low Earth Orbit. The graveyard was a free-for-all, and they were willing to look the other way, but they're not so easy in LEO. Don't forget, it's illegal to salvage junk satellites without a license, and they are cracking down on that. One operation thought a satellite they grabbed was a piece of junk. Turned out to be a weather satellite still in operation. The U.S. government was pissed, and the two scavengers went to jail. So watch it!"

The two brothers laugh. "Nothing will happen to us. We're smarter than that."

These, of course, were famous last words.

The scow ship scouts 500 miles above Earth, tracking an old satellite, catching it with a cable, and pulling it back into the ship's compartment.

Just when they are ready to leave, they are hailed by a ship of the Space Watch.

"Halt!"

The two pilots freeze.

"You're under arrest for violation of Article VIII of the Outer Space Treaty, theft of private property in space. You are to report to Base Alpha for questioning."

The two are taken to Base Alpha International Space Police Force Station for questioning.

"Do either of you have a license to operate this ship to perform salvage operations?"

"Uh, no," replies Jake.

"Do you know that in salvaging satellites, even if they are junk, a license is required and you have to notify us, and the owners of the satellite, in advance, and that includes rocket parts, and even space debris the size of pebbles."

"C'mon, really?"

"Yes, really! You're in a lot of trouble, boys!"

The two scavengers go through the legal process but are let off. Their ship is confiscated, which gets their father mad beyond reason.

Turns out that one of the licensed businesses out on a mission had witnessed the ship perform salvage operations and got suspicious. They took the license number and called the ISPF. The ship was not listed on the registry, so the space police swooped down on the brothers.

Space salvagers fiercely compete, and do not allow strangers on their turf.

SEVEN

CONSTRUCTING HABITATS

The late Gerard O'Neill, who had been a physics professor at Princeton University, had a vision of settling space by using extraterrestrial materials to build giant, revolving space habitats. The habitats could be built in different shapes and sizes, creating customized, enjoyable environments. These would be like miniature planets, where people would dwell inside them.

Before considering the construction and features of O'Neill's spaces, it helps to know the optimistic thinking behind his work, which grew out of an assignment to his advanced, undergraduate physics students. He cited the research of sociologists related to two Arctic colonies,

...one of which was British, where they paid a lot of attention to coloring and to shapes and so on, and the other of which was American, where they paid no attention to that sort of thing at all...like a prison. The results were fantastically different. In the American colony, there was continual crime, drunkenness, poor productivity....Whereas in the British colony...even though the material constraints were the same, the real effect on people was enormously different and completely a happy one.[1]

The British had no law enforcement personnel at their Arctic colony. In contrast, the American facility had to have one policeman for every twenty workers.

FROM SCRATCH

Extracting materials from the Moon or asteroids is the first step in building habitats like O'Neill's. For now, we'll focus on the Moon since the most practical research on building habitats from scratch centers on the characteristics of the Moon.

If humans are going to remain on the lunar surface, or reside in habitats in the space above, native materials should be used in building shelters, tools, and constructing spare parts, among other things. Launching everything from Earth to the Moon involves huge expense; it makes no sense considering all the materials available to make things on the Moon. The aim is to minimize payload and maximize reliance on resources available at the site. Resources like basaltic lava sourced and processed at the site might mixed with additives and fed into a 3D printer, which would then spew layer after layer of "cement" to create a shelter.

This picture taken at the Colorado School of Mines shows a roughly three-foot-by-three-foot section of a cave-like structure being created by an ICON 3D printer approximately ten feet high and ten feet wide. The PVC pipe protruding from the side suggests how material must be displaced to make room for doors, connections to the outside, interior wiring, and other requirements for negative space. The entire process is guided by computer-aided design delivering instructions to the 3D printer for where the nozzle should extrude the building material.

Additive manufacturing, which is the official industry term for 3D printing, offers enormous possibilities for using native materials for building. The opportunity (and the complication) on the Moon and on Mars involves basalt. It's one of the desired building materials found in the regolith of both celestial bodies—but basalt is a "they" not an "it." Generally speaking, basalt is a fine-grained volcanic rock rich in iron and magnesium and composed largely of the minerals plagioclase, olivine, and pyroxene.[2] Not all basalt is created equal when it comes to building a shelter for human habitation, however.

The Pacific International Space Center for Exploration Systems (PISCES), which is funded by the state of Hawaii, conducts ongoing research to determine which compositions of basalt would help create durable habitats, as opposed to those that would end up shanties. PISCES specifically aims to determine the ideal composition for sintering—that is, subjecting the material to high heat. To make a basalt brick, tile, or wall for building or a launch pad construction, for example, a 3D printer could extrude the material, and then it would be baked by solar radiation to create a unit with a structural integrity that surpasses concrete.[3]

PISCES takes full advantage of its Hawaiian location, near many lava flows and quarries rich with different compositions of basalt. After sintering, the subsequent set of tests will determine how sintered tiles hold up under pressure through structural testing. Kyla Edison, a geology and material science technician for PISCES, explains that the test will "reveal heat capacity, thermal conductivity, porosity, and thermal expansion—all of which contribute to strength and longevity."[4]

A complementary part of the challenge of building from scratch: How to locate basalt with the ideal composition for building once we have arrived on the surface of the Moon or Mars.

O'Neill developed his concepts for building space habitats a decade before 3D printing came into being, but his vision of using native materials fits with how the printer would be an integral part of construction. In his habitats, the ecosystem would have to be imported and cultivated, however. In his vision, parks, forests, lakes, and rivers would make inhabitants from Earth feel truly at home.

If cavelike dwellings printed from basaltic lava are on one end of the from-scratch spectrum, then the O'Neill colonies are on the other.

O'Neill described his plan in a 1976 book called *The High Frontier: Human Colonies in Space*. It took shape from a project O'Neill gave his advanced physics students: "Is the surface of a planet really the right place for an expanding technological civilization?" A few years later, a *summa cum laude* science student at Princeton named Jeff Bezos, who would later become the founder of Blue Origin, had direct exposure to O'Neill and his ideas. So, although concepts for the O'Neill colonies have been around for roughly half a century, they are essentially being rehashed by Bezos in presentations about the Blue Origin view of life in space.[5]

The habitats take on three different shapes: Bernal Sphere, Cylinder, and Torus. The size of the habitats can vary, depending on the desired population. In theory, they could scale up to accommodate millions of people.

In general, the habitats would be built with an airtight shell, to be covered on the interior with dirt (slag) from the Moon or asteroids. An entire ecosystem would be planted, then buildings would follow. People would have their own private dwellings. The big difference in perspective is that the horizon would curve up instead of down. Solar power would make energy essentially free, with solar power constantly provided by the steady rotation of the habitat.

BERNAL SPHERES

The Bernal Sphere would consist of a sphere with its "ends" flattened. At these ends are mirrors to reflect the Sun. On both sides of the sphere are recessed areas for agriculture, supplying food for the population year-round. At the very end of the structure, on both ends, would be space factories, with energy from the Sun radiating on solar panels where the population would work.

The shielded habitat itself would have the environment on the interior of the sphere. In a sphere with a diameter of 1,700 feet, the circumference of the sphere at its equator would be a little over a mile. Rotating at a rate of one revolution per minute (RPM), the equator would have one Earth gravity. The gravity would decrease as you moved away from its equator until reaching its poles, where one would reach zero gravity. A sphere this size would accommodate about ten thousand people.

THE CYLINDER

Using more advanced technology than the sphere, a cylinder can be made bigger, fitting more people exponentially than the first few habitats. It might look like this: A diameter four miles wide, the length twenty miles long, with a total land area of 250 square miles, discounting the solar windows. There would be six subregions: three valleys and three arrays of windows to let the Sun shine into the habitat. Light planar mirrors above the windows would open and close, allowing the Sun to "rise" and "set," forming a day and night cycle. Each valley would be two miles wide and twenty miles long. One side of the cylinder, its axis, would always point toward the Sun.

All this land area, after covering it with soil and an ecosystem, would easily allow the population of Cleveland, Ohio, or Udon Thani, Thailand—about 400,000 people—to live in comfort.

Above the axis, but connected, would be a regular or beaded torus facing the Sun and used as an agricultural area, providing food for the population. Electric power would be underground from external power stations over cables laid when the community is built. Two of these habitats could be connected, side by side, by a cable, with opposite rotations.

THE TORUS

The Stanford Torus design is a pneumatic tube—wheel-shaped on the outer rim and a central hub on the axis, connected with six spokes. Allow the diameter to be slightly over a mile, with a circumference of three to four miles. The tube itself would be over four hundred feet in diameter.

The hub, being the physical center of the habitat, would have recreation areas and low-gravity gymnasiums and swimming pools. It is the central area where the six spokes converge, each with elevator shafts, docking ports, power cables, and heat pipes. It could serve as a place for offices, shops, and laboratories. It acts as the crossroads of the colony: All people and freight entering will dock and pass through it.

The outside pneumatic tube shape is the colony itself. As the wheel rotates, the colony is provided with Earth gravity. There would be six to ten feet of slag, like the other shaped habitats, to protect from the solar flares and other types of radiation. The environment would be shaped with rivers, lakes, parks and gardens, trees, and farms on terraces. There would be room to fit ten thousand people comfortably.

The inner third of the toroidal rim surface area consists of glass panels mounted on aluminum ribs to let in sunlight from secondary mirrors. Directly above the central hub is a large but lightweight mirror, with a diameter of over half a mile. It floats above the colony with small rockets to keep its position properly adjusted. (A cable connecting the two could be another option.) It is angled away from the Sun to reflect sunlight down onto a double ring of secondary mirrors. These are mounted in rings around the hub, supported by the spokes, to reflect sunlight into the colony. The mirrors could be tipped or tilted to reflect sunlight away. This is done twice a day during the habitat's sunrise and sunset.

Directly beneath the "south pole" or "bottom" of the hub could be a second sphere, containing a possible construction shack, where major components of a solar power satellite, space factory, spaceship, or even another space habitat are put together. It could also be an industrial area where mass catchers from lunar orbit or ore from asteroids are taken so the metals can be processed.

This torus can serve as a major industrial area, where the inhabitants work and support a major space infrastructure.

PREFABRICATED STRUCTURES

In a 2012 paper written by NASA's Human Spaceflight Architecture Team (HAT), the authors outlined not only the technical needs for a deep-space habitat, but also human considerations. Compiling all the data on space requirements, they concluded that the habitat should be a four-story structure totaling almost 10,000 cubic feet (268 cubic meters) of pressurized space.[6] They broke down the required subsystems into seven different categories:

1. Equipment (22 percent of the space)
2. Mission Operations (20 percent)
3. Spacecraft Operations (20 percent)
4. Individual Space (18 percent)
5. Group Space (12 percent)
6. Logistics and Resupply (6 percent)
7. Contingency Planning (2 percent)

The team noted in the conclusion that their work stopped prior to completing detailed designs. However, it did allow each of the subsystem teams to create their design strategies. Their careful work informed the habitat developers who tried to capture NASA's attention just a few years later.

In 2016, NASA chose six US companies to develop prototypes for deep-space habitats: Bigelow Aerospace, Boeing, Lockheed Martin, Orbital ATK, Sierra Nevada Corporation's Space Systems, and NanoRacks. A primary design criterion was that each be compatible with NASA's plans for the Lunar Gateway. NASA aims for a bare-bones version of Gateway to be launched as soon as possible so it could serve as an interim station where astronauts could transfer from their spacecraft to a lander, which would then take them to the Moon.

NASA began testing five of the prototypes in the spring of 2019 and included a concept study from NanoRacks in its evaluation. Orbital ATK, now Northrop Grumman Innovation Systems (NGIS), won the contract; it seemed to be the only company that could meet NASA's tight timeline. The competing companies' different approaches to habitat design are valuable, though, suggesting viable alternatives and in some cases complementary models.

Two of those models involve inflatable technology, two others repurpose existing equipment, and one would be assembled in space.

- Sierra Nevada Corporation's proposed habitat was called LIFE: Large Inflatable Fabric Environment. The pod was twenty-seven feet in diameter, giving it an interior area of about 10,000 cubic feet. Inside, it was composed of three floors with crew quarters, a galley, an exercise area, and places to run experiments. Their model also had something SNC called an Astro Garden, which was a wall of plants that inhabitants could use to cultivate fresh produce. SNC's inflatable design would allow it to be launched from any conventional rocket. The concept was to take it where it needed to go and then inflate it and equip the living/working quarters.
- Bigelow Airspace also offered an inflatable design but referred to it as an expandable module. The unit that preceded it, which passed a two-year test on the ISS, was called the Bigelow Expandable Activity Module (BEAM). Like Sierra Nevada's inflatable, Bigelow's could be carried by a conventional rocket. The one intended for the Gateway complex was called the B330; it had 330 cubic meters of interior space, or about 11,650 cubic feet. As with any of the habitats, in addition to normal creature comforts like a reliable zero-g toilet, it would

have radiation shielding and other protections. The exterior of the habitat was eighteen inches thick, with dozens of layers of material in that shell. Think back to the description of the *Apollo 11* spacesuit, with its twenty-two layers of material, each being "mission critical." Bigelow also has designed a smaller version called Destiny, with roughly one-third the interior area of the larger habitat.

- Lockheed Martin's habitat, which they began building at Kennedy Space Center in 2018, was a refurbished Multi-Purpose Logistics Module (MPLM). MPLMs were large pressurized containers used to transfer cargo from space shuttles to the ISS. Lockheed's efforts were with the Donatello module, which never made it into space. It measured 22 feet long (6.7 meters) and 15 feet wide (4.57 meters), with anticipated use by astronauts not exceeding sixty days. Like the others already described, it included racks for scientific investigations, life support systems, sleep compartments, recreation facilities, and workstations.

- Boeing's idea was to launch their habitat in four parts and then assemble it so it could be an integral part of the Gateway intended to orbit the Moon. Boeing's experience in constructing space stations—the company is the prime contractor for ISS—gave it a unique background for designing a habitat. The idea was to build on what worked on the space station in developing a module.

- The winning concept offered by Northrop Grumman was a habitat small enough to launch on its Cygnus automated cargo spacecraft, which delivers cargo to the ISS. Despite its compact size, it's designed for comfort.[7] The fact that Cygnus already has a track record with NASA gave the agency some assurance that the NGIS habitat could be delivered by 2020.

- NanoRacks came into the competition with a unique plan to repurpose spent launch vehicle upper stages. Its habitat was called Ixion and the design emerged after a five-month NASA-supported study concluded that it was possible to convert an Atlas 5 upper stage left in LEO into a habitat module. The NanoRacks concept involved refitting the abandoned upper stage with life support and other systems similar to what other companies were proposing. NanoRacks was the only company among the six that did not involve development of a ground prototype.

None of the designs were intended for use on the lunar surface, but they all suggest an approach to a shelter that could be planted on the Moon.

In 2019, NASA funded two habitat-development "institutes" through Space Technology Research Institute (STRI) grants, which involve a collaboration of industry and university partners. One was the Resilient ExtraTerrestrial Habitats Institute (RETHi).

> RETHi seeks to design and operate resilient deep space habitats that can adapt, absorb, and rapidly recover from expected and unexpected disruptions. The institute plans to leverage expertise in civil infrastructure with advanced technology fields such as modular and autonomous robotics and hybrid simulation.
>
> Through an integrated effort, RETHi will mature deep space habitats that can operate in both crewed and uncrewed configurations. The institute plans to create a cyber-physical prototype testbed of physical and virtual models to develop, deploy, and validate different capabilities.[8]

Finally, there is something called HOME: Habitats Optimized for Missions of Exploration, a NASA-funded multi-university Space Technology Research Institute.[9] The current commitment involves $15 million over five years; the funds are STRI grants. The HOME team includes:

- The University of California, Davis
- The University of Colorado, Boulder
- Carnegie Mellon University
- The Georgia Institute of Technology
- Howard University
- Texas A&M University
- The University of Southern California
- Sierra Nevada Corporation
- Blue Origin
- United Technology Aerospace Systems

A primary concern is ensuring that autonomous systems in the habitat perform at peak efficiency. Therefore the team includes experts in machine learning and robotic systems, in addition to a host of engineers.

The lead investigator at Carnegie Mellon University is Mario Bergés, a professor of civil engineering, who sees a prime role for civil engineers in space:

> Since the beginning, civil engineers have been the stewards of the infrastructure that supports modern life. If humanity is moving into space, it makes sense for civil engineers to be part of that.[10]

THE PERSONAL TOUCH

At the Center for Intelligent Environments (CENTIENTS), an interdisciplinary team of students and faculty at the University of Southern California is doing research on bidirectional interactions between humans and buildings. The group includes scholars and innovators from engineering, computer and data science, social science, design, and other fields related to human-centered design and artificial intelligence. With support from the National Science Foundation, they have established projects to prove their premise:

> Recent advances in artificial intelligence have opened doors for opportunities to integrate human behavior and preferences with automation to develop personalized, dynamic work and home environments.[11]

Although CENTIENTS does not focus specifically on habitats designed for space settlements, it takes little imagination to see how easily the advantages would translate if introduced while those shelters were taking shape on the Moon or Mars. The human-centered approach to building could make off-Earth habitats become environments that give settlers a sense of control and familiarity. Pioneers could leave Earth and go home.

PART IV

THE ADVENTURE

EIGHT

SPACE TOURISM

The argument for space tourism is simple: Space tourism would pave the way for humanity to settle space. The theory is that, through tourism:

- The cost of space transportation would lower exponentially.
- Space transportation would improve and diversify.
- Space habitation would be improved.
- Safety would increase in all areas.

Despite all the good reasons for it, however, some within the space community still mock the concept of space tourism. The rhetoric has softened a bit as more business plans are backed up by technological advances. It's rare to see something as severe as the opinion column in *Wired* magazine referring to it as a "millionaire boondoggle thrill ride" after the fatal crash of *SpaceShipTwo* in October 2014.[1] The journey for would-be space tourists has been fraught with financial loses and dangerous failures. However, the reasons for pursuing it have not disappeared.

From a financial perspective, suggesting the investment of billions in a space tourism venture is folly belies reality. The World Travel and Tourism Council estimates that over 260 million people worldwide are employed in the tourism industry.[2] Annualized employment growth in the industry in the United States, which has almost six million people in the industry, is 1.7 percent. Worldwide spending on travel exceeds $950 billion. Tourism accounts for 11 percent of the global domestic product. Combined travel and tourism exceeds $1 trillion, being the world's largest industry.

There are tours up Mount Everest and to the deep sea in submersibles; the heavens are next for people with money and a taste for adventure. Space tourism is on its way up.

PRECURSORS TO SPACE TOURISM

Space camps feature motion-based space simulations and activities related to space exploration. Since they are geared toward people with a range of fitness levels and a broad range of ages, they nurture a taste for space, but not preparedness for it.

The next step up is zero-gravity flight. The American KC-135, also known as the "vomit comet," has a maneuver where the plane flies up and then makes a nose dive, simulating zero gravity. The plane flies in the form of a parabola between the altitudes of 24,000 and 32,000 feet. The maneuver is similar to that of a roller coaster. The plane flies nose high up into a 45-degree angle, then over a peak and then descends, producing zero gravity as it flies downward at 45 degrees. Everything inside the plane is weightless, floating around at zero-g. As the plane levels off at 30 degrees, nose up, gravity returns, allowing participants to gently settle on the aircraft floor. Finally, the g-force is increased slowly to about 1.8 Gs until the aircraft reaches an altitude of 24,000 feet, and then the maneuver is repeated. The tourists get to float around for a few minutes, or however long it takes for the plane to do its maneuvers until the "tour" ends.

NASA accommodates tourists periodically at places such as the Glenn Zero Gravity Research Facility, although the NASA facility is primarily devoted to ground-based microgravity research. The Glenn facility is a large shaft—a 510-foot-deep tunnel known as a "drop tower" that is evacuated to eliminate air resistance, allowing for about five seconds of weightlessness.[3]

The Russian equivalent to the vomit comet is a modified version of the Ilyushin II-76, the IL-76 MDK, and slots aboard it can be booked through Space Adventures, Inc. Just like the vomit comet, it makes a series of parabolic maneuvers, with each offering a zero-gravity experience for about twenty-five seconds.

Peter Diamandis, founder of the X Prize, also offers zero-g rides on a Boeing 727 specially modified for his company, Zero Gravity Corp. The aircraft performs parabolic arcs to create a weightless environment, with a single ride starting at $5,400 per person. The Zero Gravity website indicates that ride includes "15 parabolas, your own ZERO-G flight suit, ZERO-G merchandise, a Regravitation Celebration, certificate of weightless completion, photos and video of your unique experience."[4] An impressive list of endorsers includes astronaut Dr. Buzz Aldrin, the late renowned theoretical astrophysicist and cosmologist Professor Stephen Hawking, and culinary and lifestyle maven Martha Stewart.

A CHECKERED HISTORY

The space tourism industry has run the gamut from the ridiculous to the sublime, with the ridiculous generally having more to do with timeline than concept.

Starting with the earliest success in this business, Space Adventures has brokered deals for super-wealthy tourists with Russia's space agency, Roscosmos, since the turn of the century.[5] Between April 2001 and September 2009, seven different civilians rode a Soyuz spacecraft to the International Space Station at a cost of between $20 million and $35 million each. One of those ISS tourists was Richard Garriott, a gaming software magnate and founder of Space Adventures. Space Adventures currently employs people who train tourist astronauts on the ground, plan and market flight experiences, and forge deals with public and private sector entities for slots on their spacecraft.

After five of Garriott's clients had visited ISS, the Barcelona-based space tourism company Galactic Suite Design announced its Galactic Suite Project.[6] The long range plan involved an "orbital hotel chain with modular space accommodation . . . (and) the design and position in relation to the equator will allow visitors to orbit Earth fifteen times every day—and to see fifteen sunrises!"[7] The first resort on an orbiting shuttle was scheduled to open within five years of the announcement, that is, by 2012. That timeline was later extended, with a test launch planned for 2014. They didn't meet that target either.

Galactic Suite's ambitious timeline failed to match its technological preparation. The company never was able to acquire a rocket system for transport, much less equip an orbiting craft with the amenities of a resort facility. In 2008, Galactic intended to start charging €3 million ($4.4 million at the time) to reserve a three-day stay in space, with the fee also covering training at a luxury hotel on a Caribbean island. Unlike some companies, such as XCOR, they did not collect the money in advance.

XCOR had a plan for early space tourism that earned respect. As Space News later declared: "XCOR Aerospace, for a time, appeared to be the biggest contender to provide suborbital space tourism."[8] Using a spaceplane taking off horizontally, rather than launching vertically, one pilot would take one passenger past the Kármán line, hang out for a few minutes so the rider got a taste of space, and then come back to Earth, landing on a conventional runway. The initial ticket on XCOR's Lynx spaceplane was $100,000 per flight and they thought they could do five flights a day.[9] Unfortunately, XCOR's exquisite technology and spaceplane design were not supported by a cohesive business plan. Their scheme for space tourism helped split the company's focus and drive it into bankruptcy. Tourists were left with the old-fashioned vertical launch and a trip to the International Space Station.

In the meantime, aspiring space tourists had already sent XCOR their $35,000 deposits (or in some cases full payment), with great hope of experiencing ninety seconds of weightlessness around 2011 or 2012.[10] The date kept slipping, with paying customers looking at 2015 or 2016. And then XCOR declared bankruptcy in November 2017. An estimated 282 people had already committed funds for their ticket.[11] People who invested that kind of money were not focused on their financial loss; more than anything, they were disappointed.

The canceled XCOR flight was devastating for ticketholder Cyril Bennis, who dreamed of going to space since meeting NASA astronaut John Glenn as a child. Bennis had even prepared himself for the suborbital flight by visiting the National Aerospace Training and Research Center in Southampton, Pennsylvania.[12]

As Art Thompson, co-founder of Sage Cheshire Aerospace says, "In my opinion, companies use space tourism to attract investors and customers because 'everybody' wants to be an astronaut."[13]

Another company with a spirited idea sucked in strong interest from would-be space tourists because of their affiliation with NASA astronauts and senior personnel. In this case, the "tourists" were intended to be government representatives rather than private individuals. The Golden Spike Company aimed to develop a "cislunar superhighway."[14] Cislunar refers to an area between Earth and the Moon or the Moon's orbit. They intended to fly crews to the Moon and back by 2020 for an estimated $1.4 billion per flight. In Golden Spike's short run from 2010 to 2013, the company managed to attract some extremely high profile and respected people to its board. They included venture capitalist Esther Dyson; former presidential candidate Newt Gingrich; Jonathan Clark, M.D., a former NASA Space Shuttle flight surgeon; the widow of *Apollo 12* moonwalker Peter Conrad; the acclaimed space historian and author Andrew Chaikin; and as president and CEO, former NASA science chief Dr. Alan Stern.[15]

PRESENT AND NEAR FUTURE

Like the Golden Spike Company, SpaceX has its eyes on the Moon. Although Elon Musk's original timetable was off—he had announced he would send two tourists on a trip orbiting the Moon in 2018—he has a track record of launches suggesting he can make lunar tourism for a select few a reality by the early 2020s.

The really exciting news is this, though: Lots of ordinary rich people are headed into space soon.

On October 28, 2019, Richard Branson rang the opening bell at the New York Stock Exchange. The joyful noise announced the dawn of space tourism from a business perspective. Virgin Galactic had just become the first commercial space flight company to list on the stock market. For $250,000 for a 2.5-hour ride, adventurers could book a suborbital flight with Virgin Galactic. More than six hundred people had handed over their money by the time Branson rang the bell, with Branson aiming to send people into space every thirty-two hours by

2023.[16] Branson sees the company moving from a loss position in 2020, to break even the following year, and a projected $274 million in earnings by 2023.[17]

Branson has invested in Reusable Launch Vehicle (RLV) technology that launches horizontally for his space tourism venture. Despite a disastrous accident in 2014, the spaceplane design and launch concept that Virgin relies on has proven to be sound—more that fifteen years of R&D went into it before Branson confidently rang the Stock Exchange bell.

After Burt Rutan's company, Scaled Composites, won the Ansari X Prize in 2004 with the flight of the Rutan-designed *SpaceShipOne*, Branson saw the potential of it. A vehicle that could leave from and return to a conventional runway struck him as well suited for tourism. He funded the Spaceship Company to work with Scaled Composites to build *SpaceShipTwo*. Unfortunately, after successfully performing its first test flight in 2013, *SpaceShipTwo* broke up in a test flight the following year and crashed in the Mojave desert. Its successor, VSS *Unity*, has been successfully doing test flights since late 2018.

Competition at the relatively low price point of $250,000 is expected to come from the China Academy of Launch Vehicle Technology. Although, by the time the Chinese venture is up and running, Virgin Galactic may have at least a five-year head start. The China Academy plan involves a suborbital journey, focused on a view of the heavens and the experience of weightlessness, enabled by technology developed by China Aerospace Science and Technology Corp. They will also use RLVs with the appearance of fixed-wing aircraft, but the venture is not expected to enter service until 2028.[18]

Blue Origin is on roughly the same timetable as Virgin Galactic for getting tourists into space, but it has a more traditional approach—that is, a vertical launch. It has a decidedly less traditional approach to landing, however: also vertical. As of mid-2019, the suborbital rocket known as *New Shepard*—named after astronaut Alan Shepard Jr., credited with being the first American to travel into space—had launched successfully ten times. It had been in development for thirteen years at that point.

The ride on *New Shepard* will deliver a stunning view given the capsule's six giant windows. About four minutes of weightlessness, and then a gentle trip back to Earth with the capsule under a parachute and the trip will be over. Blue Origin has not said how much it will cost, nor has the company sold seats yet.

Boeing is an elephant in the industry, and when it forged an agreement with NASA on commercial development, it marched forward with a commanding presence. Boeing's CST-100 Starliner began as a crew capsule, but the agreement with NASA has always allowed for the inclusion of paying space tourists on Starliner.

Boeing has produced an inviting video aimed at those tourists who want a ride on the Starliner:

> Welcome to the all-new Boeing Starliner. Humanity's space taxi. All designed to be as lightweight as possible. The 20 engines and 28 thrusters of your autonomous spacecraft have enough power to reposition the space station into a higher orbit and avoid speeding space debris. Able to fly to and from the international space station up to 10 times with up to seven passengers or a mix of crew and cargo. The Starliner is the first U.S. capsule to return astronauts on land rather than the ocean.[19]

Boeing adds some comforting features to Starliner's offerings—features that suggest what space tourists paying an enormous fee would expect: "wireless internet and tablet technology."[20]

Near-term plans for space tourism also focus on destinations, not simply a ride to space. On September 12, 2019, Bigelow Aerospace announced that it had hosted NASA personnel for a two-week ground test of a space station called the B330. Although the announcement indicated the autonomous expandable independent space station was destined for Mars, it has also been envisioned as part of an orbiting hotel complex. Moving into the space hotel business is a natural for

Bigelow Aerospace, whose founder, Robert Bigelow, made his fortune by founding the Budget Suites of America chain of hotels.

The two modules under construction, B330-1 and B330-2, would link together to create the space hotel and offer double the cubic capacity of the ISS.[21] (The B330s got their name from having 330 cubic meters of volume.) They could operate in LEO and cislunar space, with other modules potentially being added later to create a giant orbiting station. Bigelow has not affixed a price to the cost of the six beds on board a B330. However, estimates run in the millions of dollars.

Clear evidence of a viable offering in the marketplace is the presence of competition. Orion Span has plans for a commercial space station that would host about six space tourists at a time, just like a B330.[22] Orion Span has an interesting hook: Don't just be a tourist, be a volunteer researcher. Their tourism model involves a partnership with universities and research organizations flying experiments onboard their spacecraft, called *Aurora Station*. Training to support the scientific research would be part of the pre-flight training for flights beginning around 2022.

OUT THERE

Touring the Moon would begin to develop soon after we return there. At first, luxury cruisers would go into lunar orbit and return to Earth, similar to the scenario proposed by Golden Spike. Soon after, a hotel on the Moon would be built—perhaps a Bigelow expandable habitat with recreational facilities. One adventure would be to suit up and explore the surface, either by walking or by a Lunar rover. Dune buggy races around the Moon, or guided tours of Apollo landing sites would be on the agenda. In Chapter 9, we look at the real prospect of a Lunar museum. And in the final chapter, we look at visions—pinned to specific technologies such as those developing at SpaceX—of routine transport to Mars.

MAKING IT LEGAL

In the United States, a law was enacted in 2004 addressing the privileges and constraints of space tourism. The European Union also has laws governing commercial space transportation, and other countries have developed laws or guidelines as their space tourism industries develop.

To give an example of how a government grapples with the legal implications of an emerging industry, we'll look at the US effort.

In response to the coming space tourism of suborbital flights, the US Congress passed, and President George W. Bush signed into law, the Commercial Space Launch Amendments Act of 2004 (H.R. 5382). This authorized the Office of Commercial Space Transportation, a branch of the Federal Aviation Administration (FAA) to grant permits to private spaceship operators hoping to fly passengers into space. This means that any passenger who is not part of the crew (that is, a space tourist) must be informed of the risks of space travel before liftoff. Passengers fly at their own risk, just as a skydiver acknowledges through signing a waiver that any jump out of a drop zone's aircraft is an accepted risk.

The terms of the law were temporary, lasting until 2012, when the FAA issued regulations for crew and passenger safety. This eight-year grace period (2004–2012) was to help promote commercial travel by placing the industry on a firm regulatory footing, allowing it to evolve and develop new technologies to make journeys into space as common as flights overseas. Standard regulations then took over, including safety regulations for the licensed space vehicle, along with providing safety features for the participants.[23] Any vehicle carrying passengers into space does require licensing and insurance, along with adherence to safety regulations.

NINE

SETTLING THE MOON

s the Moon just a rock in the creek we step on to get to the other side—to Mars?

Robert Zubrin, founder of the Mars Society and a highly regarded aerospace engineer, sees both the Moon and Mars as destinations. In his book, *The Case for Space*, he offers a detailed plan for a lunar base and he calls Mars "our new world."[1]

In a July 12, 2012, interview, NASA administrator Jim Bridenstine seemed to downplay the importance of the Moon in plans for space exploration: "We need to keep our eyes on the horizon goal. The goal is not the Moon. The goal is, in fact, Mars."[2] He sidestepped that sentiment a week later during a ceremony in the Oval Office celebrating the fiftieth anniversary of the Apollo Moon landing. In response to President Donald Trump's repeated question about going to Mars directly from Earth, Bridenstine called the Moon a proving ground before going to Mars because "we need to learn how to live and work on another world."[3]

Even if we went from Earth straight to Mars, at some point, humans will likely learn to live and work on the Moon. And when we do, science educator Bill Nye cautions that we should talk about settling on the Lunar surface rather than colonizing the Moon.[4] Nye earned widespread fame as TV's "Science Guy." He heads the not-for-profit Planetary Society, founded in 1980 by Carl Sagan, Bruce Murray, and Louis Friedman—all godfathers of people who feel their place is in space. Nye's preference for settlement may sound like a lurch toward political correctness; however, it simply makes more sense. Settlement suggests community. Colony suggests outpost—something far from civilization. The space community Nye represents hopes that we will bring civilization and an appreciation for community with us to Lunar settlements rather than leave those traits behind.

Establishment of a lunar settlement will involve different functions, all interdependent upon one another: Scientific investigation of the Moon and outer space and research for new space technologies; development of lunar resources; and the building of a self-sufficient and self-supporting lunar base, evolving into a lunar city. These are the key step-by-step processes in developing the Moon as a place where humans remain for long periods of time—and perhaps call home.

WHY ESTABLISH AN INHABITED MOON BASE?

As Jim Bridenstine said at the fiftieth anniversary celebration for *Apollo 11*'s landing, the Moon is a proving ground. Humans have learned to inhabit "alien" environments on Earth such as the Arctic, however, those environments have breathable air, organics, and other substances people need to live. Learning to live and work on the Moon is solid preparation for the challenges of functioning on a planet quite different from Earth that is millions of miles from us.

The establishment of an inhabited lunar base has been proposed for three different functions:

- Scientific investigation of the Moon and the application to research problems
- Exploitation of lunar resources for space-based industries
- Development of a self-sufficient and self-supporting Lunar settlement

The first proposal seemed the only one likely to reach implementation when NASA first planned a lunar base. Years ago, it would have involved governments funding all scientific research without expecting a payback, in either money or industrial technology, for the foreseeable future. The dynamics of scientific investigation changed when space companies and universities began putting company-sponsored experiments on the ISS.

The second set of activities, which could help fund research efforts, is exploitation of Lunar resources by private corporations in partnership with government agencies. It's a necessary step both in terms of revenue generation and extracting and processing the raw materials to build on the Lunar surface and maintain equipment.

The third aim of a Lunar settlement would logically be realized as the first two get established. People associated with research and with resource operations would use their knowledge and locally sourced materials to complement whatever they brought with them from Earth. They would make the settlement scalable and independent of Earth, with an ever-expanding population base.

A Lunar base built for a self-supporting settlement is, as Bridenstine suggested, a model for spending any length of time on Mars. With a self-supporting lunar base, near term achievements will include:

- The first practical experience toward adapting to life on another planetary body
- The first demonstration of humanity's ability to alter the environment of space through innovative use of native materials
- Development of infrastructure to tap natural resources of space to create profitable enterprises
- Demonstrating much of the technology and possibly acquiring some resources necessary to support subsequent human exploration of Mars
- Advancing physical and life sciences to create new views of our universe and to help us understand the space environment

BEFORE WE BUILD

Simulations on Earth will provide the essential scientific and technical knowledge to succeed on the Moon.

The Lunar test bed at the Colorado School of Mines looks like a sandbox filled with gray dust with some coarse material, rocks, and a crater. Tracks over the surface illustrate how a mini rover tried to traverse the simulated regolith. The experiments indicate how contamination from dust might affect the machinery, potential mobility issues, and the kind of forces that will be needed to drill in different areas. The data collected then informs those designing systems for navigating and drilling on the Moon.

Students use the Lunar test bed, but companies also come to Mines to test their equipment. Lunar Outpost, founded by a Mines graduate in 2017, is one of those companies that both use the lab facility and bring students onboard as interns and employees. On its website (lunaroutpost.com) the company has a two-minute video that shows how a Lunar Outpost Resource Prospector—a rover—is attempting to move through and over the different surfaces of a test bed like the one at Mines. It spotlights features such as 3D-printed wheels, all-wheel-drive, autonomous navigation, differential suspension, and 360-degree LiDAR (Light Detection and Ranging) to demonstrate how it's possible for the vehicle to navigate, map, and inspect Lunar terrain. And then the "prospector" part of the name takes center stage when a subsurface sampling drill makes its debut and the vehicle conducts sampling and analysis of material. The entire machine weighs only 22 pounds (10 kilograms).

The plan is to send not just one rover, but rather a swarm of them to collect enough data identify the type of resources in a given area, as well as their location, depth, and so on. They would communicate with each other, aggregating information and "deciding" where to go next to make a comprehensive sweep of the territory.

The prospecting challenge is that patches of the Lunar surface may be homogenous for a long stretch and then be very different where an asteroid had struck and left remnants of iron, nickel, and other minerals. There is no erosion or humidity to distribute the material, so it stays where the impact occurred. Volcanic eruptions also have affected the homogeneity of the surface by creating basaltic plains. These large dark areas that Johannes Kepler observed in the seventeenth century and called "mare" were so named because he thought they were oceans. In contrast, the landscape of Earth lacks homogeneity because of water and wind creating erosion and movement. Whatever happens on Earth gets stirred up by natural forces.

At Mines, testing excavation systems is part of work that goes with identifying resources and drilling methods. The challenge to developing those systems is the unusual nature of Lunar regolith. On Earth, there is fine dust—pebbles turned into small particles by water and wind. On the Moon, without humidity or any other source of erosion, the particles are the result of being pulverized. A microscopic view of them reveals jagged edges, which makes them extremely abrasive. Mines's director of the Center for Space Resources,

Dr. Abbud-Madrid, explains, "If they got into your lungs, they would destroy them in no time. They do the same kind of thing to machinery."[5]

The upper few centimeters of regolith are fluffy. Below that, the dirt gets very dense. Every time the Moon has been struck or experienced a lunar quake, when the dust settles and remains in place, it eventually compacts. As of this writing (we repeat this phrase occasionally because discoveries are logged daily), the Lunar surface has only been examined to a depth of less than two meters. During the *Apollo 15* mission in the summer of 1971, astronaut David Scott debuted the Lunar Surface Drill and tried to extract a subsurface core. The first 40 centimeters (about 16 inches) were easy, but from then on, he struggled. After 1.6 meters (just over 5 feet), he got stuck. Eventually he was able to pull the drill out and then moved to the next hole. He only got to a depth of one meter at the second location, so the drilling exercise was terminated.

Ground studies to develop drills for lunar excavation involve blocks developed to replicate the different forces likely to be encountered on the Moon. The result will be having an array of drills carefully matched with the material.

BUILDING ON THE MOON

As we described in Chapters 1 and Chapter 7, Lunar rocks can be crushed and mixed with cement paste to form concrete useful in building Lunar structures. Early Moon inhabitants may rely on habitats designed and built on Earth, but as time goes on, an army of 3D printers that have ingested local concrete could build a neighborhood.

In addition to minerals, glass is an abundant material found in Lunar soil. On Earth, water combining with elements composing the glass weakens the material. The absence of water on the Moon, along with the vacuum, makes Lunar glass extremely high in tensile strength, competing with metals.

With processing, glass can be separated from the Lunar soil to form fibers, slabs, tubes, and rods. Fibers can be useful in tensile stress situations, such as bulkheads in a space habitat or beams and columns in a space structure. Mixed with metals (for example, iron) or concrete, fibers can aid in Lunar construction. Tunnels can be lined with Lunar glass to seal in air. Other glass products can be also be produced, such as satellite components, fiber optic cables, and lenses.

Lunar glass is strong enough to be substituted for structural metals in a variety of space engineering applications, both on the Moon and in LEO. This will enhance economic utilization of the Moon, and save launch costs of space station and Lunar habitat components that would otherwise have to be launched from Earth, by manufacturing them on the Moon. This will also add to the Moon's growing economy, with glass becoming another export item, besides oxygen, regolith, and possibly concrete.

In the minerals area, besides oxygen, glass, and helium-3, the lunar soil also has ample supplies of the following major minerals, in the following form:

Silicon. . . . . (SiO_2)

Sodium. . . . .(Na_2O)

Iron. . . . . $(FeO; Fe_2O_3)$

Potassium. . . . .(K_2O)

Aluminum. . . . . (Al_2O_3)

Manganese. . . . .(MnO)

Magnesium . . . .(MgO)

Chromium . . . . (Cr_2O_3)

Titanium. . . . . (TiO_2)

Calcium . . . . (CaO)

Usable amounts of nickel-iron alloy.

As indicated by the chemical designations, oxygen is combined with all these minerals.

At first, these materials can be used for experimentation, to test the technical feasibility for product development. These experiments would be done by chemists, chemical engineers, and metallurgists. Processing regolith, separating minerals, and then processing them would involve either adapting terrestrial technology to the Moon's environment, or devising fresh technologies.

One form of new technology to be applied to lunar minerals is powdered materials processing, or sintering, as applied to ceramics. The powdered metal is then injected into a mold or passed through a die to form a weakly cohesive structure near the true dimension of the intended object to be manufactured. The product is then formed by applying pressure, high temperature, long settling times, or a combination thereof.

Powdered metal processing offers the following advantages:

- It is good for making alloys.
- It allows fabrication of products from starting materials that would otherwise decompose or disintegrate.
- It is more flexible than case, extrusion forming, or forging.
- Solar energy can be used in the process.
- Sintering can produce lunar bricks for construction, and ceramic products, such as superconductors, which would be of great value on Earth, especially for transmitting and saving energy.

In addition to the major elements of the Lunar soil there are trace elements. These trace elements can be used to produce the water, oxygen, fatty acids, vitamins, and sugars that are needed for life support systems, as well as plastics. Trace elements include:

Hydrogen...(H)
Helium..(He)
Sulphur...(S)
Neon..(Ne)
Phosphorous...(P)
Argon...(Ar)
Carbon..(C)
Krypton...(Kr)
Nitrogen..(N)
Xenon...(Xe)

Platinum Group Metals (PGMs) are another group of trace or residual elements in the Lunar soil. These are the metals considered particularly valuable in the off-Earth mining ventures on asteroids and the Moon. They include:

Platinum..(Pt)
Gallium...(Ga)
Osmium..(Os)
Germanium...(Ge)
Iridium...(Ir)
Chromium..(Cr)
Ruthenium...(Ru)
Zinc..(Zn)
Gold..(Au)
Paladium..(Pd)

A pilot plant would be the first to process Lunar minerals. At first, small components would be manufactured, such as silicon chips and photovoltaic cells. As Lunar manufacturing advanced to making larger and more complex products, large manufacturing plants could be built. Among the products would be:

- Habitats for the Moon, space stations, and, eventually, Mars
- Propulsion systems
- Fuel cells
- Construction materials
- Space vessels
- Superconductors, made from ceramics and alloys
- Shielding material
- Components for solar power satellites (SPSs) for Earth
- Other ceramic and metal products, large and small, for export, adding to the Moon's economy

SETTLING IN

In earlier chapters of the book, we described some of the infrastructure elements needed for Moon settlement, namely habitats and ground transportation. Anyone familiar with NASA's proposed Lunar Orbit Gateway probably has detected that we see an accelerated pace toward prospecting for resources, building on the Moon, and settling as minimizing the appeal of the Gateway.

What we endorse is more akin to Robert Zubrin's plan for a Lunar base, as expressed in his book *The Case for Space*. Compared to the proposed Gateway, it is potentially a cost-effective, revenue-driven program for developing the Moon and settling on it. At the same time that occurs, the push to Mars can continue, perhaps with greater efficiency and fewer costs borne by governments.

Zubrin's no-Gateway scenario begins with humans spending a couple of months at time in temporary quarters on the Lunar surface and working with robotics to mine and process resources, principally water ice to create propellants and a power supply for fuel cells. It's a step that reduces the cost of subsequent missions because propellant for space vehicles is available onsite. As a corollary, it delivers other resources that would be building blocks of the Lunar base.

The astronauts get there by catching a ride on a commercial launch vehicle like the SpaceX Dragon, which carried their Lunar Excursion Vehicle (LEV) as part of the payload. They would hop in the LEV for the first part of the return trip, rendezvous with a Dragon once again, and come back to Earth.

Each successive mission would extend the duration of the trip, incrementally building the base. Exploration missions on the surface would also lead to outposts.

The base would start out by having cargo landers deliver space-station-derived facilities for habitation. Power could come from Kilopower, a small nuclear reactor developed by NASA; the concept is to have something transportable great distances before providing ten kilowatts of power for ten years.[6] NASA completed successful tests of the Kilopower Reactor Using Stirling Technology (KRUSTY) in March 2018. Robots like NASA's ATHLETE Rover could lift facilities off of the cargo landers and then assemble them for the base.

What this scenario illuminates is the use of existing, tested NASA technologies in conjunction with commercially developed space vehicles and systems to set up operations on the Lunar surface. It also takes advantage of native materials to enable building, exploration, and repairs.

Lunar soil, about two meters deep, is one option to cover the habitats, for shielding against solar and cosmic radiation, especially against solar flares. Shielding with water could also be used. Perhaps six people in each discrete facility would now have a setup for daily living, with oxygen and energy. The Bigelow B330 is built to accommodate six people for a tourism experience or four people for a long-term venture, so this kind of inflatable or expandable habitat could be a building block for the first settlement.

Next would come the development of agricultural facilities. This would be the first long-term life support system, involving the regeneration of air, water, and the production of food. Hydrogen is one of the main elements that would have to be imported for years to come, for lack of hydrogen on the Moon. (Note: Hydrogen does exist in the lunar soil, but, like helium-3, it originates from the solar wind.)

Eventually, the base would be a self-supporting facility with long-term life support systems, agricultural research and production, resource extraction and processing facilities, power stations, and an experiential knowledge base about human functions and relationships. (Sociologists and psychologists take note: There is a place for you on the Moon and beyond.)

The evolution of a self-supporting Lunar base would probably take decades. In the meantime, it would facilitate trips to Mars and advance human experience in living and working in an off-Earth environment.

GOING TO WORK ON THE MOON

We begin with opposites. At least on the surface.

Two very different, early twenty-first-century Lunar careers captivate the imagination. The first sounds like an old-fashioned, low-paying job for an adjunct professor at a university: historic preservationist. The second demands an advanced knowledge of microbiology with a strong command of geology: space bio-miner. After zeroing in on those, we look at how the Moon hosts other ways to make a living for individuals and engender financial success for companies.

WHAT'S OLD IS NEW

Although countries other than the United States might not care about heritage sites on the Moon as much as the United States, the six sites where Apollo crewman touched down are "sacred ground" to some Americans. The White House Office of Science and Technology Policy (OSTP) issued a report in 2018 called "Protecting and Preserving Apollo Program Lunar Landing Sites and Artifacts." Among other topics, it presented the legal underpinning for ensuring the sites of original footprints and rover tracks are left undisturbed.

In his book, *Moon Rush*, Leonard David offers an elegant answer to dilemma that involves more imagination than national or international law:

> There may be scientific and engineering value in revisiting and perhaps even removing and returning hardware to Earth from some of the other sites at some point in the future. Those plans need to incorporate the concerns of historians, archaeologists, and others about the cultural and historical nature of the locations. Arguably the best solution could be a museum on the Moon. Tourists could see firsthand how the human exploration of our neighboring worlds began. A first step is preserving the Apollo sites now, so that the opportunity still exists in the future.[7]

If David's vision comes true, in a decade or two (if we proceed at a measured pace, as is customary for academics), we will see job openings on the Moon for archivists, conservators, curators, docents, exhibit designers, and historians. And of course, every major museum needs good managers, security personnel, and someone to run the gift shop.

In a special digital way, a museum on the Moon already exists. A tiny ceramic wafer (1.9 cm by 1.3 cm [0.75" by 0.5"]) was supposedly planted on the Lunar surface in November 1969 after being secretly attached to the *Apollo 12* Lunar module *Intrepid*. Thanks to the etching technology of Bell Laboratories, it features art contributions from John Chamberlain, Forrest Myers, David Novros, Claes Oldenburg, Robert Rauschenberg, and Andy Warhol. In theory, it's up there at Mare Cognitum, waiting to sprout successors.

JOBS FOR HUNGRY BACTERIA

Space biomining uses bacteria to extract resources from regolith, defined as loose, heterogenous material covering solid rock. We could call it Moon dust or Mars dirt.

Iron may not sound as precious as platinum or gold, however, it is crucial to sustaining life on Earth, and it has different forms, one of which is considered common and one of which is considered rare. A certain type of bacteria called *exoelectrogenic bacteria* likes eating iron. Then what they excrete delivers the desired form of iron which can easily be separated from the rest of the material. It's a symbiotic relationship: Give the bacteria something they want, which is iron III (ferric iron or hematite, which is common), and they excrete iron II (ferrous oxide, which is rare). Who are these magical, metal processing organisms? They are microorganisms with the ability to transfer electrons extracellularly.

While the jobs on center stage belong to exoelectrogen bacteria, the human jobs belong to those with combined expertise in geology and microbiology. This sounds like a new space-related hybrid, but it really isn't. Dr. Zea estimates that that 5 to 15 percent of the gold and copper we use today on Earth already comes from biomining.[8]

REAL MONEY ON THE MOON

Every time the level of complexity of an experiment escalates, with movement toward the final product, the work must be done in a relevant environment. When the Moon is the relevant environment, companies and their partners at universities and NASA can make a strong case about their desire to do R&D on the Moon. It would be similar, if not identical, to the processes dominating decisionmaking in countries other than the United States. To put this imperative in context, it's helpful to look at a technology progress chart developed by the US Defense Department that describes the Technology Readiness Level (TRL) of a product.[9] TRL is a system to gauge technology maturity that is based on a scale of 1 to 9, with 9 meaning that it's ready for market.

1. Basic principles observed and reported
Lowest level of technology readiness. Scientific research begins to be translated into applied research and development. Examples might include paper studies of a technology's basic properties.
2. Technology concept and/or application formulated.
Invention begins. Once basic principles are observed, practical applications can be invented. Applications are speculative and there may be no proof or detailed analysis to support the assumptions. Examples are limited to analytic studies.
3. Analytical and experimental critical function and/or characteristic proof of concept.
Active research and development is initiated. This includes analytical studies and laboratory studies to physically validate analytical predictions of separate elements of the technology. Examples include components that are not yet integrated or representative.
4. Component and/or breadboard validation in laboratory environment.
Basic technological components are integrated to establish that they will work together. This is relatively "low fidelity" compared to the eventual system. Examples include integration of "ad hoc" hardware in the laboratory.

(continues)

5. Component and/or breadboard validation in relevant environment.
Fidelity of breadboard technology increases significantly. The basic technological components are integrated with reasonably realistic supporting elements so it can be tested in a simulated environment.
6. System/subsystem model or prototype demonstration in a relevant environment.
Representative model or prototype system, which is well beyond that of TRL 5, is tested in a relevant environment. Represents a major step up in a technology's demonstrated readiness.
7. System prototype demonstration in an operational environment.
Prototype near, or at, planned operational system. Represents a major step up from TRL 6, requiring demonstration of an actual system prototype in an operational environment such as an aircraft, vehicle, or space.
8. Actual system completed and qualified through test and demonstration.
Technology has been proven to work in its final form and under expected conditions. In almost all cases, this TRL represents the end of true system development. Examples include developmental test and evaluation of the system in its intended weapon system to determine if it meets design specifications.
9. Actual system proven through successful mission operations.
Actual application of the technology in its final form and under mission conditions, such as those encountered in operational test and evaluation. Examples include using the system under operational mission conditions.

If we want to increase the TRL of a technology destined to be used in space, we need to be able to continue testing beyond Earth: first on a space station to assess the microgravity aspect, and then on lunar surface to assess the partial gravity aspect and the conditions that this would operate under. The result is a product that actually works on the Moon, and possibly beyond the Moon. Research can go only so far on Earth with certain products and processes. After that, testing and tweaking to take them to a high TRL must be done in an off-Earth environment—that is, an environment where mission conditions are present.

Anticipating that the TRL steps to product deployment will remain a valid standard for work on the Moon and other space environments, NASA produced a handbook called *Microgravity-Based Commercial Opportunities for Material Sciences and Life Sciences: A Silicon Valley Perspective*. The authors associated with NASA Ames Research Center focused on how the first four of the nine steps are a guide to assessing "the potential of microgravity for public benefits and economic growth over the next decade."[10] The book was published in May 2014 so the vision was that projects that met the progressive evaluation criteria would be headed to stages 5 through 9 in the 2020s. In other words, they would be bound for space and ultimately remain in space for use there or be shipped back to Earth to improve conditions and processes here.

The authors summarized the first four criteria as follows, and logged a few conclusions that infuse the process with optimism:

1. **Potential.** "Products from space have evident superior technical performance."[11]
2. **Credibility.** Successes emerged from previous investigations, but attracting commercial interest means more must be done.
3. **Accessibility and Awareness.** Space is perceived as high-risk, so commercial researchers tend to shy away

from doing investigations in microgravity. Companies and university researchers need a searchable, dynamically updated database focused on commercial work in microgravity.

4. **Interest.** "Identifying industry specific infusion points for microgravity driven discoveries is key."[12]

In trying to entice companies to plan for Lunar and space station R&D, the Ames team cited compelling examples of high TRL products welcomed by the market. Among the results on the list were:

- Bulk metallic glasses that were used in myriad ways, including golf clubs, surgical tools, and a SIM eject tool for the iPhone.
- Semiconductor crystals used for fabrication of low noise field effect transistors (FETs) and analog switch integrated circuits (LCS).
- Capillary flow investigations yielding software for modeling of complex interface configurations and a new rapid diagnostic process for infant HIV.

Product and process advances such as these that were already on the market before 2020 illustrate the real and immediate opportunities for making money on the Moon.

ABLE-BODIED: A WHOLE NEW MEANING IN SPACE

Microgravity or zero-gravity environments are not just beneficial for R&D related to products and discoveries. They also present work opportunities for people well-suited to those environments. Aretech has done research, published in the *Journal of NeuroEngineering and Rehabilitation*, documenting mobility of a person with complete paralysis in both legs in a zero-g environment. Aretech's system detected signals from the person's brain, and then sent those signals to electrodes at his knees to create movement.[13]

With current research like this in mind, consider the following story about Bob working on the Moon.

Bob was a paraplegic. During a college football game, his spine was broken. Medical technology had come a long way by that time, but it still had not gotten to point of completely repairing spinal injuries.

The accident did not deter him from his love of sports, nor did it keep him away from the dance floor. His wheelchair tore it up!

Even with his great spirit and lust for life, Bob faced a major frustration as he neared graduation. He had studied civil engineering and had always wanted to spend his days outside, on construction sites for bridges and buildings. Two of his closest high school buddies wanted the same life and they were just about to get it. Not so for Bob.

Up in Earth-Moon space, work had just begun on a metal-processing factory, where resources mined from asteroids and the Moon would be mixed to form new alloys in near-zero gravity, making them stronger than those formed in Earth's gravity. Superconductors, hard cutting tools, and hard metals more resistant to the extreme temperatures of space were examples of what was needed, and in great demand by other space manufacturers. However, they needed laborers and young professionals willing to spend years developing the plant, and it would involve some strenuous work. They would have to spend their time there in near zero-g.

The forewoman, Jean Claudia, made her statement at an executive meeting. "I don't want workers coming up here to do a two-year stint and leaving again, with us in a situation where we have to waste time and money training new people every couple of years. You know, these college students who want 'experience' in space before going to their ivory towers. I want dedicated workers who are willing to spend their careers here so we can grab hold of the market."

One of the staff replied, "You know, this is a zero-gravity environment and people are used to Earth gravity."

"And...?" replied Claudia.

"People aren't going to waste away their bodies for you."

"We can't have one-g facilities here. You know that. You seem to be doing well!"

"That's because I was disabled back on Earth. I have mobility advantages here in zero-g that I didn't have down there. Here, I can move all of my body. Why don't you recruit more people like me? They can spend the rest of their lives up here working on construction projects, research, or whatever."

"And they can wear spacesuits and move heavy loads. People in a wheelchair on Earth can be roughnecks up here."

Bob saw the ad:

Are you wheelchair bound? Work with us!
Use your whole body in outer space.

Live a productive life, a daring life, a life
most people would only dream about living.
Ace Metals has the right job for you!

Bob applied and was later called in for a job interview in Minneapolis, Minnesota. When he arrived, there were many applicants, wheelchair bound, brace covered, lining up to work in space.

And work they did.

Many other space factories between Earth orbit and the Moon caught on. Metals, chemicals, construction workers, satellite repairmen and women, many wheelchair-bound people from all over Earth countries flocked to these zero-gravity factories where they would willingly spend the next few decades of their lives.

TEN

SETTLING MARS

Long before we settle on Mars, we have to make preparations that differ greatly from, but build on, planning for a Moon community. While our coverage of transportation in space and on the ground as well as habitat construction still applies to a Mars experience, there are unique and complex challenges related to human health and well-being.

On the Moon, people are 238,900 miles (384,400 kilometers) away from their Earth home. On Mars, they are anywhere from 34.8 million miles (56 million kilometers) to 250 million miles (401 million kilometers) from Earth. The shorter distance, or launch window to send a spacecraft to Mars, occurs roughly every two years and two months.

The distance factor plays a major role in the challenges of travel to Mars and settlement of it. As a corollary, while some characteristics of the Red Planet are inviting in terms of habitation, most are not. The distance from Earth will make it difficult to meet survival and protection needs, such as food and medical care, without regular supply runs.

WHY MARS? WHY HUMANS?

Different people answer this question with either aspirational or scientific reasons. Going back to the Introduction, the biggest aspirational motive is that putting people on Mars lifts us up. The accomplishment itself—just like Neil Armstrong setting foot on the Moon—inspires curiosity, innovation, and confidence in our ability to solve problems and achieve great feats.

Scientific reasons for going deliver an equally strong dose of inspiration for many of us. Dr. Joel Levine, who participated in half a dozen successful missions to Mars and is a forty-one-year veteran of NASA, cites two major scientific reasons for sending human beings to Mars:

1. Determine whether there is past or present life on Mars.
2. Learn why Mars experienced cataclysmic climate change.

If we find life on Mars today and if we can show it's indigenous life— life that formed on Mars—it will make tremendous advances in our understanding of biochemistry and molecular structure of life and it will provide information on the human body and how to treat diseases that we haven't even thought about.[1]

The second reason could tell us if Earth is destined for the same fate as Mars. Discoveries on Mars reveal it was a very different place in its early history. Formed 4.6 billion years ago at the same time Earth took shape, it had characteristics such as Earth does today—a thick atmosphere, lakes, and rivers. It also had huge ocean at least a mile deep that covered most of the planet's northern hemisphere. The features of Mars have been confirmed through decades of having rovers roaming the surface and high-definition images from sources such as the European Space Agency Mars Express satellite. Today, Mars has ice craters, but no detectable surface water. It's surrounded by a very thin atmosphere. Somehow, it became an inhospitable place.

Ice cores taken from the Artic, the Antarctic, and glacial ice have told a vivid story about climate changes on Earth. We need to do the same kind of analysis with ice cores from Mars to find out its story. Levine explains that the bubbles in ice cores contain essential elements of the narrative: "In these ice cores, there are bubbles, and in the bubbles are trapped atmosphere—not (from) today, but when the ice was formed millions and millions of years ago."[2] Retrieval of an ice core taken from a thousand feet deep could then lead to discovery of critical information regarding the climatic history of Mars.

Levine is very clear that tasks such as hunting for signs of life and extracting ice cores from ideal locations are too complicated for robotic missions. Human beings must be involved in these investigations for a few intertwined reasons. First, robotic missions still have to be preprogrammed. Our robots do not have a level of artificial intelligence enabling them to identify the significance of anomalies and investigate further. Second, robots at the 2020s stage of development don't incorporate human ingenuity or physical agility.

Levine also cites "speed and efficiency" as a human differentiator, and that may come as a surprise to some. As supporting evidence, he noted that a well-known Mars scientist told him he can do in two hours what it takes a rover six months to do. It's a matter of having intellectual agility. Human researchers can observe, extrapolate, analyze, make connections, and then jump back to observation and analysis, all the while rejigging their investigative criteria as they learn.

The answer to "Why humans?" in the context of research and exploration seems clear, but the extended question relevant to this chapter is "Why should humans settle on Mars?" We would assert there is no "should." It's a matter of desire.

GETTING THERE

Not to be glib, but we know how to get to Mars. We have the technology, although there remains some controversy about which technology is best suited for the job. And if Mars One CEO Bas Lansdorp is correct, the big technological problem is not getting there; it's getting back to Earth.[3] Before looking at his vision for a Mars one-way trip, let's consider more conventional approaches.

At a space exploration conference in Guadalajara in September 2016, Elon Musk premiered a video promoting what SpaceX then called the Interplanetary Transport System (ITS). Instead of projecting a sense of animated science fiction, the video presents a scenario that has an engaging believability to it. After viewing it and hearing Musk's presentation on ITS, science writer Rebecca Boyle wrote in the *Atlantic*, "Even among tech companies, whose product announcements are geared to be grandiose, Elon Musk's Mars-colonization rollout feels like something new."[4]

The video shows a line of people walking from the top of a silo, across a bridge to a spacecraft poised to launch vertically. The nose of the launch vehicle appears to be windows in a latticework. A caption identifies the launch location as Pad 39A, the *Apollo 11* Launch Pad at Cape Canaveral, Florida. After stage separation, the spacecraft soars toward parking orbit and the booster heads back to Earth, landing neatly on the launch mount where it stood for lift off. The silo releases an arm, which drops a chain that tethers a propellant tanker; the arm swivels around to load the tanker onto the booster.

The video then shows the in-orbit refueling of the spacecraft, with the tanker returning to Earth and the spacecraft zooming toward Mars. The craft's speed during its interplanetary coast is 100,800 km/hr or 62,634 mph. Solar arrays deploy, providing a source of continuing power throughout the journey. With the Red Planet in sight, the spacecraft prepares for Mars entry with the temperature now reaching 1,700 degrees C (3,092 degrees F). Landing gear is deployed and the door opens to reveal a stunning, if somewhat barren, landscape with the planet then spinning and morphing into something looking more like Earth. Mars is terraformed in seconds.

Musk's ITS daydream was inspiring, if not completely doable, and he seemed to realize that and refined the plan. By October 2019, he unveiled Starship, the successor to ITS, although it had briefly gone by the name Big Falcon Rocket, or BFR. Unfortunately, it blew up during a pressure test the following month. Nonetheless, he does hope to get to Mars by the mid-2020s.

NASA wants to get to Mars as well, but the space agency is not on the same timetable as SpaceX. The plan is to get a crewed vessel in low-Mars orbit by the early 2030s and land on the Red Planet sometime after that.

Boeing CEO Dennis Muilenburg pushed back at Musk's claim that SpaceX would lead the way, despite the difference in timetables between NASA and Musk's company. Boeing is the main contractor for the first stage of NASA's Space Launch System for deep space missions, and Muilenburg insisted to CNBC's Jim Cramer that "the first person that sets foot on Mars will get there on a Boeing rocket."[5]

Among the other NASA prime contractors with a key role in going to Mars, Lockheed has stepped up with plan for a Base Camp for the first crewed mission. The Base Camp is a system composed of the NASA's Orion deep-space crew capsule, the Space Launch System, habitats for the crew, and solar electric propulsion. Lockheed has built redundancy into every life-support, energy-generation, and communication component of the Base Camp. The plan supports a thousand-day mission to deploy rovers and drones to investigate the surface and collect samples; explore Mars's moons, Phobos and Deimos; and do site selection for the first human landing. The company also has plans to be a central player in the crewed mission that lands on Mars and is developing a Mars Lander that uses Orion avionics and systems.

Finally, for those who want to run away from home and toward new adventure forever, Mars One will help them. Dutch Entrepreneur Bas Lansdorp wanted to go to Mars ever since he saw a picture of the Mars Pathfinder rover named *Sojourner* in 1997. He was twenty years old at the time. After founding and selling his shares in a successful power company, he turned his full attention to reaching Mars and funded Mars One. Originally, Lansdorp hoped the company would make a profit, but when he announced plans for the Mars colony, offers of volunteer support and donations poured in. Lansdorp responded by turning Mars One into a nonprofit organization.

His original timetable included sending a rover on a recon mission in 2016, following that with missions to drop off supplies, and then taking people to Mars by 2023. That has since slipped to 2031. Every couple of years after, consistent with the time when Earth and Mars are closest, Mars One would shuttle people to the surface and leave them there with the transport vehicle that also carried supplies.

Lansdorp, who has no plans to develop his own technology, estimates that it will cost at least $6 billion to send the first four people to Mars and about $4 billion for each group of four after that. Sounds like a hefty price tag, but the Mars One founder has proven on more than one occasion that he is a savvy entrepreneur. At a TedxDelft presentation, he explained to the audience:

> The Olympic Games in London generated $3 to $4 billion dollars in revenue just from TV broadcasting rights and sponsorship. That's just three weeks of broadcasting.

Mars One is a not-for-profit organization, but we will still have to finance this mission…and we want to do this by offering the entire world as an audience for this mission.[6]

The first episodes of the show would feature the selection process for the first mission. Candidates will train for the mission at a copy of the Mars outpost on Earth. The expectation is that billions will watch the "competition" for a slot as well as the moment of selection. Cameras will also catch the homesteaders when they arrive, and as they acclimate.

Lansdorp believes his plan is not only a way of financing the ventures, but it also achieves the more meaningful purpose of involving the whole world in the creation of a Mars settlement.

THE HUMAN FACTOR

Conditions and practices on Earth make us ill-suited for a journey to Mars followed by settlement on an inhospitable, distant planet. First, we live in a cocoon, shielded from harmful radiation by a protective blanket of atmosphere. Second, we go down the street to a doctor for routine checkups and to a doctor or hospital when the body malfunctions.

On the way to Mars and on Mars, these things we take for granted will be unavailable. The mandate for Mars settlement means coming up with different ways to safeguard our life and health.

PROTECTION FROM RADIATION

Astronauts on the ISS have exposure to radiation that a physicist from the Belgian Nuclear Research Center measured at 100 to 150 times higher than on Earth.[7] Radiation exposure has an impact on human health, but the Center found that not every human experiences the same impact. Different people apparently have varying cellular capacity to repair the DNA damage caused by radiation. Some people simply have a higher radiation tolerance and seem to walk away from the ISS experience and regain excellent health.

The radiation unit at the Center, which is headed by Dr. Sarah Baatout, evaluates the potential risks of ionizing radiation on health. Her team also researches ways to protect against the harmful effects of radiation. Baatout and colleagues in the field have run the numbers on the differences in radiation exposure on Earth, the ISS, the Moon, and Mars:[8]

- Exposure on Earth is the baseline, with the atmosphere shielding us from all but about .1 percent of cosmic radiation.
- ISS exposure is 100 to 150 times greater than on Earth.
- On the Moon, the exposure is 300 to 400 times greater than on Earth.
- The trip to Mars would, at times, deliver about 1,000 times more radiation than on Earth.
- On the surface of Mars, exposure would be about 2.5 times greater than the ISS, according to data collection by the Mars Odyssey probe.[9]

In other words, the biggest threat to safety is posed by the trip, not the destination. The weight of shielding materials known today makes blanketing a spacecraft in them impracticable, so the concept is to restrict enhanced shielding to certain parts of the ship. That doesn't solve all the problems, however.

Of the three types of radiation that Baatout says will blast the spacecraft carrying people to Mars, one is relatively inconsequential, one is predictable, and one involves serious concerns for humans. They are:

- **Solar wind.** This is a continuous flux of low energy particles deterred by the spacecraft's skin.
- **Solar particle events.** These are storms involving high energy particles. The intermittent flux of radiation can be deadly, but the event is predictable by minutes or even hours. With this advanced notice, passengers to Mars can go to a protected part of the spacecraft. In addition, there are ways to reinforce personal protection and human resistance to the effects of the ionizing radiation.
- **Galactic cosmic rays (GCRs)** present a different class of problem. We don't yet have completely efficient shielding to stop them, according to Baatout, although NASA and the Center have some ideas.

NASA has as an official statement on GCRs and how we can deal with them when we transport people to Mars:

GCRs [are] particles accelerated to near the speed of light that shoot into our solar system from other stars in the Milky Way or even other galaxies. Like solar particles, galactic cosmic rays are mostly protons. However, some of them are heavier elements, ranging from helium up to the heaviest elements. These more energetic particles can knock apart atoms in the material they strike, such as in the astronaut, the metal walls of a spacecraft, habitat, or vehicle, causing sub-atomic particles to shower into the structure. This secondary radiation, as it is known, can reach a dangerous level.

There are two ways to shield from these higher-energy particles and their secondary radiation: use a lot more mass of traditional spacecraft materials, or use more efficient shielding materials.[10]

NASA, contractors, and partners in other government space agencies are therefore investigating shielding materials to protect against galactic cosmic radiation. In 2018, they also received substantial help from some high school students in Durham County, North Carolina.

The students speculated that shields composed of radiation-eating mold could be effective in protecting space travelers and developed their theory after learning that *Cladosporium sphaerospermum* has thrived at the site of the 1986 Chernobyl nuclear disaster. Led by Graham Shunk, who entered North Carolina School of Science and Mathematics after graduating from high school in 2019, the students got help from Kentucky-based Space Tango (see Chapter 6) in sending samples of the mold to the ISS in December 2018.

The result was an automated CubeLab investigation that utilized the growth of the mold to explore its potential to create a radiation barrier. The Higher Orbits student team hypothesized that the fungus would thrive in a spaceflight environment due to increased radiation exposure, and, since the fungus feeds off of radiation through radiosynthesis, it may create a radiation barrier. Radiosynthesis is similar to photosynthesis, except that the organism converts radiation into chemical energy instead of converting sunlight.

Space Tango's CubeLab Engineer—Biological Systems and Laboratory Manager Andrew Diddle worked alongside the student team to design and build the CubeLab. When the mold got to the ISS, astronauts put it into Petri dishes, leaving half of each dish empty. Every 110 seconds after that, for thirty days, Geiger counters measured radiation levels beneath the dishes. The students could claim victory when the results came in: the counters measured a 2.4 percent decrease in average radiation levels underneath the mold-cover portion of the Petri dishes.

Shunk, backed up by other researchers, theorizes that "if the mold were about 21 centimeters thick, it could provide humans adequate protection from radiation levels" on their exploratory mission to Mars.[11]

Having already learned from ISS astronaut/cosmonaut studies that certain people seem to have a high tolerance for radiation, she suggested that this knowledge could affect astronaut selection, but it could also lead to genetic modification of people aiming to travel to Mars. She notes "twenty beautiful biomarkers"[12] that seem to protect an individual against ionizing radiation. Following this premise to the next level, we might speculate that the people best suited for Mars settlement are those genetically well suited to handle radiation—either naturally or through enhancement.

Another solution could be pharmacological compounds to boost the body's resistance to radiation. Microbiologists, physicians, organic chemists, pharmacologists—professionals in many different fields have a role in determining how to make this a reality.

And then, there's hibernation. The Center's researchers determined that hibernating animals, with their slowed metabolism for prolonged sleep, have an increased radiation tolerance. Bear biologists have a role in our space exploration program.

Conclusion: Put at least some Marsbound people to sleep for an extended period, give them a drug to boost their resistance to radiation, genetically enhance them if necessary, and put superb shielding on the spacecraft. We are not trying to be simplistic here. These are schemes to mitigate problems we need to examine seriously before we send human beings to Mars.

PRECISION MEDICINE

Jay Sanders, the physician known as "The Father of Telemedicine," began working with NASA on Mars-related medical challenges in 1998. He calls the type of care required to sustain and restore the health of Mars travelers "precision medicine"—fundamentally, it means treating every person as a unique being rather than as a member of a homogenous population.

On its face, that premise seems indisputably logical, however, that is not common practice for healthcare on Earth. For example, normal blood pressure is standardized at 120/80 or less. The numbers refer to the force of blood against the artery walls while the heart pumps blood to the body (systolic pressure) over the force of blood against the artery walls and the heart relaxes and refills with blood (dyastolic pressure). If an individual's personal normal is 90/60, then a 120/80 reading is high; it signals an aberration. The person whose routine physical yields "good numbers" could have hypertension.

Baseline numbers related to the entire spectrum of medical tests are needed for any astronauts, so precision medicine is practiced to a limited degree already. Sanders presents a vision to take that further, so that there would be an enhanced predictive quality to test results.

Precision medicine necessitates knowledge of genetics. The discipline goes beyond the knowledge of genetic makeup, however. Gene function can be affected by environmental factors; examining how that occurs and the expression of the gene changes in subsequent generations is epigenetics. People exposed to gravitational changes, radiation, and the other physiological stresses related to space travel will probably not—quite literally—be the same people they were when they left Earth.

Tracking the changes to determine if cardiovascular, immunologic, or any other body system problems are evolving must occur efficiently in this practice of precision medicine for Mars travelers. That calls to mind two questions: How will the tracking be done? Once the data are available, who—or what—takes the steps necessary to address the emergent medical issue?

As for the "how," Sanders proposes a couple of ways: "Nano sensors. They would be less than seven microns, which is the diameter of a red blood cell. They would be circulating along with your red blood cells and function as an artificial sensing capability."[13]

Another possibility is accessing and analyzing interstitial fluid, that is, the fluid around the body's cells. Its job is bringing oxygen and nutrients to the cells and remove waste products from them. And the simple explanation of why it's valuable in diagnostics is that it can tell stories about DNA. Sanders sees the use of microneedles as a way to draw the fluid, and while that is a known minimally invasive approach, it had its critics. In October 2019, however, a Chinese team studying DNA related to Epstein-Barr Virus from interstitial fluid published its findings on a hydrogel microneedle patch they developed. It seems very promising in terms of speed and accuracy of results.[14]

With diagnostic data available, the ship's surgeon can step in. But as we said in Chapter 2, history has taught us that the practice of medicine in remote locations holds unique challenges. In early seafaring days, the ship's surgeon was often someone with minimal training trying to provide care he was unqualified to give. On modern battlefields, a skilled anesthesiologist might be on the scene, but what's needed is an internist; the only way to render appropriate care is with guidance from a specialist through a telemedicine system.

Telemedicine is not a viable option en route to Mars or once people arrive. Signal latency means that by the time a problem presents and is communicated to the appropriate physician on Earth or a space station, fifteen or twenty minutes have elapsed. Same time lag with the response. If the problem is acute, the patient could die.

Ideally, the ship's surgeon is an autonomous, AI-equipped surgical robot assisted by humans with some medical training. The status and diagnostic information collected by sensors and microneedles feeds Dr. Robot's database, keeping it updated at all times. Unquestionably, there would be communication of that data to the Mars medical team on Earth, but Dr. Robot would take whatever action is required on an urgent basis. In a time of crisis, signal latency makes it absurd to think that the robot arms could be manipulated from Earth to perform an operation.

One or two robots like this—redundancy is important in space—not only augment human abilities, but they also restore a degree of predictability of outcomes that humans find very reassuring.

Sanders looks at Mars travel as an opportunity for everyone on Earth: "I view our ability to go to Mars as solving most of our problems here on Earth in terms of the healthcare delivery system."[15]

TERRAFORMING FOR THE 3000S

The vision of terraforming the Red Planet—that is, turning it into an Earthlike environment to support human life—has given rise to fascinating science fiction as well as engaging stories that pretend to be rooted in science. As we noted above, Mars started out like Earth, with oceans and rivers and a thick atmosphere. It hasn't been like that for a long time, though. And based on the technology we know now, it won't be that again for a long time. This statement is not meant to discourage, but rather to inform Marsbound pioneers that taking your food and down comforters with you is going to be de rigueur for quite a while—like a thousand years or more.

We would like this prediction of a terraforming timeline to serve as a challenge: Prove it wrong. Create the technology that advances terraforming because, if you do, then you can concurrently benefit millions of people on Earth who are stuck living next to deserts they can't inhabit or farm.

Here's a scenario offering inspiration. Joel Levine, referenced before as a forty-one-year veteran of NASA exploration including six successful Mars missions, offers this sequence for ultimately terraforming Mars.[16] A key word is "ultimately," but in a way, it's up to you figure out how far in the future this really is:

1. **Build a large solar reflector in space to direct solar radiation down to heat the planet's surface.** The objective is to release massive amounts of frozen water and CO_2. Carbon dioxide (CO_2) is a greenhouse gas. We have too much of it now on Earth and therefore have

many activists screaming that we have to reduce CO_2 emissions to mitigate climate change. If we want promote climate change on Mars to warm it up, however, then engendering the production of it serves our purposes.

2. **Seed the Martian surface with photosynthetic plants to convert the CO_2 to oxygen, which we breathe.** What kind of plants qualify? Well, anything green is a good start.

3. **Nurture an ozone layer.** With oxygen becoming more prevalent in the atmosphere, a natural chemical response from incoming ultraviolet radiation from the Sun is formation of an ozone layer. At this point, the protection from solar radiation that we have experienced in the cocoon atmosphere of Earth can be realized on Mars.

4. **Sustain the emergence of liquid water on the planet.** A warm and thick atmosphere hospitable to humans will mean melting of the crater ice and other sources of frozen water on Mars. We have to be stewards of the oceans and lakes.

Settling on Mars is great and beautiful challenge. Everything we need to do to make it work is something we can and should do now on Earth.

CONCLUSION

Space Is Our Neighborhood

Some people look to the heavens from a young age and dream of shooting straight up, flying, and touching celestial bodies. Some people have career trajectories and interests that lead them, logically, toward space. For both types, space becomes part of what we consider our "neighborhood"—the area surrounding us that comes to feel familiar.

Right now, it's still mostly a rough neighborhood. It takes people with certain skills and resilience to operate in this 'hood, but those people show us how thriving in space-related careers is achievable.

In many ways, the journey of Angel Abbud-Madrid embodies the aspirations of so many people we talked with about joining space research efforts at universities, companies, and within governments. It began with dreams and curiosity, with curiosity keeping dreams alive despite what other people said was possible.

Growing up in Mexico during the Apollo years, he dreamed of working in a space field, but the United States and the Soviet Union dominated during the space race. He saw no opportunity to fulfill his dream until he came to the United States for his graduate degree in engineering. He met a professor who said, "Do you want to get involved in space-related work?" Abbud-Madrid jumped at the chance.

"What can I do?" he asked.

"We'll get you involved in a project simulating low gravity Earth."

Taking experiments to the space station was expensive, so the objective was to do them on Earth as cost-effectively as possible. Simulating pressure and temperature can be done without too much difficulty, but simulating gravity is a different matter. At the time, Abbud-Madrid had to do his experiments in a drop tower, offering five seconds of no gravity. But he needed more time. The choice was a taller building or going into an airplane that would ascend to 30,000 feet and then drop abruptly by 10,000 feet. During that fall, he could observe his experiments with no effects of gravity. He flew not dozens, but perhaps more than a thousand of those drops in pursuit of a career in space.

"I threw up, but you get used to it."

He later had opportunities to have his experiments onboard the shuttle and Space Station.

Angel Abbud-Madrid is now director of the Center for Space Resources at the Colorado School of Mines. From Chihuahua to New Jersey to space—a great, long journey.[1]

Unlike Abbud-Madrid, Jean Wright did not set out to have a career in space. She began sewing at the age of seven.[2] When she grew up, she became a dressmaker. Then in 2005, she saw an article in her local paper about a team at Florida's Kennedy Space Center that had a vital role in protecting astronauts. Housed in the Thermal

Protection System Facility at the Center, they stitched the thermal protection for the space shuttle. The amount of hand-sewing that went into forming fiberglass and ceramic-based fabrics into lifesaving panels surprised her—and it woke her up to an opportunity of a lifetime.

She went from seamstress at a dress shop to Aerospace Composite Technician, joining a group of eighteen who called themselves "The Sew Sisters." They built and repaired thermal protection flight hardware and parachutes for the space shuttle and Orion.

> The more we could place blankets instead of tiles on the outside, the lighter the shuttle could be. We saved an enormous amount of weight with those layers, allowing the shuttle to use less fuel and carry heavier payloads such as satellites and science experiments...[3]

The team also fabricated large heat-shield blankets to encircle the engines. Attaching those blankets for flight involved hand-stitching with a metallic, nickel-based, and temperature-tolerant superalloy called Iconel 625.

Wright did this thermal work for six and a half years, until the very last shuttle mission. Although it may sound tedious, it involved creativity as well as meticulousness—staying tuned into the science and mechanics that affected her work.

In the case of both Angel Abbud-Madrid and Jean Wright, the demands of their space "neighborhood" included unwavering commitment to precision and mental agility—the curiosity and desire to look for the best way to get the job done.

The foundation of scientific methodology supports an ability to think through problems and reconfigure approaches to solutions as requirements change. The foundation of creative methodology supports communication and common understanding while those rapid shifts are taking place.

We need Angel Abbud-Madrid and Jean Wright, as well as people in the arts, humanities and trades, to make space a neighborhood where we feel connected and directed.

In the end, the desire for space exploration is all about what we enjoy on Earth. The job opportunities, commercial development, expansion of scientific knowledge to improve living conditions and health—they can all make life better on Earth. Let's also not lose sight of the way space inspires dreams and stories. Both the scientists and the poets among us are an important part of our adventures in space.

APPENDIX A

Sampling of Space-Related Jobs and Careers Referenced

Aerospace composite
   technician
Aerospace engineer
Archeologist
Astronaut
Avionics engineer
Avionics technician
Bear biologist
Chemical engineer
Civil engineer
Computer scientist
Conservationist
Cosmochemist
Curator
Energy manager
Ergonomic designer
Exhibit designer
Geneticist
Geologist
Heavy equipment operator
Horticulturist

Machine learning engineer
Mechanical engineer
Metallurgist
Microbiologist
Nanotechnology engineer
Nuclear physicist
Organic chemist
Pharmacologist
Physician
Pilot
Public service professional
Radiologist
Remote sensing scientist
Robotics designer
Robotics operator
Sewing professional
Software engineer
Social scientist
Space debris collector
Space suit assembler
Telemedicine coordinator

Thousands of current positions open in the space industries worldwide
are listed on the Space Talent website: https://www.spacetalent.org/.

GLOSSARY

Additive manufacturing. 3D printing.

Bioastronautics. Study of the effects of space flight on living organisms.

Biomining. Use of bacteria to extract resources from regolith.

Carbonaceous asteroid. Asteroid containing a lot of water in the rock.

Chondrules. Tiny sphere embedded in meteorites that some scientists think are the building blocks of planets and moons.

Cislunar. Area between Earth and the Moon or the Moon's orbit.

Cosmochemist. One who explores the mystery of chondrules and other chemical compositions of the universe, and the changes taking place.

Galactic Cosmic Rays (GCRs). Particles accelerated to near the speed of light that shoot into our solar system from other stars in the Milky Way or other galaxies.

In Situ Resource Utilization (ISRU). Collecting, processing, and using native resources instead of importing them from Earth.

Ionizing Radiation. Radiation with enough energy to cause ionization in the body or material through which it passes.

Kármán Line. An imaginary line in and around Earth's atmosphere, at 330,000 feet (6.2 miles or 100 km) above sea level that is considered the starting point of outer space.

LaGrange Libration Point. A point between two heavenly bodies (for example, the Earth and the Moon) where the gravitational forces from both bodies cancel each other out, permitting any object at these points to stay in place.

Micrometeorite. Particle ranging from fifty micrometers to two millimeters that survived entry through Earth's atmosphere, usually found in sediments.

Optical mining. Drilling with light.

Platinum Group Metals. A group of metals, usually platinum, ruthenium, palladium, osmonium, rhodium, and iridium, often found in the same mineral deposits.

Regolith. Layer of loose material covering solid rock; used to describe the material covering the Lunar and Mars surfaces.

Remote Sensing. Use of satellite focused on Earth or other moons and planets examining their features.

Signal Latency. Period of delay between when an audio signal is input and when it is picked up on the other end.

Sintering. Exposing material to high heat to make it coalesce, similar to baking a cake.

Solar Particle Events. Storms involving high-energy particles.

Solar Wind. Continuous flux of low-energy particles.

Stratosphere. The second layer of Earth's atmosphere, between the Troposphere and the Mesosphere, extending from six miles (10 km) above sea level to thirty-two miles (50 km).

Superconductors. Metals offering little or no resistance to electricity.

Technology Readiness Level (TRL). System to gauge technology maturity that is based on a scale of 1 to 9, with 9 meaning it is ready for market.

Telemedicine. Communication between medical personnel in different locations, often for diagnostic or treatment purposes.

Terraforming. Turning a region into an Earth-like environment to support human life

Volatiles. Nitrogen, water, carbon dioxide, ammonia, hydrogen, methane, and sulfur dioxide: ingredients needed to help sustain human life, found on both the Moon and Mars.

ACKNOWLEDGMENTS

ALASTAIR

I never intended to be original. Many of these opinions, recommendations, and so on, were teachings and advice from others whom I did not mention in the citations and the chapters themselves, because there are too many facts to fit everyone in. But I would like to mention them here, in appreciation for all they have taught me.

The late David C. Webb, founder of the Space Studies program at the University of North Dakota (UND), located in Grand Forks (and still going strong), for his many recommendations on how we should pursue space in ways that most of us would never have imagined.

Joanne Irene Gabrynowicz, a space law professor first at UND, and then at the University of Mississippi (Ole Miss) from whom I learned a lot about law, and life, and who remains a very good friend. The other professors in the Space Studies department at UND: James Webb, Dick Parker, Grady Blount, and last, but not least, Chuck Wood.

To then US senator Byron Dorgan (D-ND) who received copies of my book to give to important officials back in 1990, and who wrote me back saying "you have a great future ahead of you, Alastair." I haven't failed you, Senator Dorgan.

Charlie Shaw and Chad Glass, who did illustrations for me at my request.

To the National Space Society, where I've worked and where the staff recommended that I apply to UND Space Studies, and to all I've met at the many International Space Development Conferences that I've attended through the years.

To the late Charles Sheffield, who told David Webb about me at a conference in LA (I wasn't there, that was the beauty of it.)

To Kathy Fick and Tim Megorden, deacons at the Christus Rex Lutheran Center, and also to Nancie Ziemke and Rand Rasmussen, then keepers of the Rex, where I spent many long hours composing the first draft of this book, back in 1990.

To my classmates at the University of North Dakota, where I spent many happy years.

To my parents, Howard and Doris Browne, who funded me while I worked on this book throughout the years.

Chris Turner and Rachel Maloney, who offered me much useful advice on writing this book.

Robert Godwin, of Apogee Books.

Jeff Krukin, Rick Tumlinson, and Will Watson, who woke me up to the fact that space will be developed by private enterprise, not any government.

To the late Eddie Geschwanter, who in a conversation in Touro Park in Newport, Rhode Island, unwittingly restored a dream of mine about humanity going to space, which stayed with me to this day.

Pat Dunfey, a schoolmate of mine who once said to me that J. R. R. Tolkien ... "knew that he couldn't have dragons or elves, so he did the next best thing. He put it on paper." Obviously, I followed through on that statement.

James Clifford Hobbins, my high school history teacher who always knew I could do it.

Lucia Baker Owen, my English teacher in my junior year of high school, who taught the ethics of space exploration in a class covering the literature of fantasy.

Susan Sperling Tingly who said that I would be a success. (You'll get the first copy, autographed, I promise.)

Salvatore Ribera, Cathie DeCesare, and Michael Mathieu, who predicted that this book will sell big.

Robert Zubrin, of the Mars Society, on his publishing advice.

To all whom I've given copies through the years, such as Michael Griffin (then head of NASA), and Ray Bradbury.

Last, but not least, Maryann Karinch, who was not only my agent, but also my co-author.

All those who joined my Facebook page, "Alastair Browne on Space Development," I came through for you.

There are many people I have met through the years, whose names are too many to remember, but I appreciate anyway in encouraging me to persist until this book becomes (hopefully) a success. You know who you are.

Thank you all.

MARYANN

Although Alastair Browne and I did not connect until 2017, he and I had both started researching topics in this book decades before. My first thanks go to Alastair for contacting me, collaborating with enthusiasm, and sharing his knowledge, research, and sources.

At the heart of this book are the experts, in addition to Alastair, who generously shared their time, expertise, and amazing stories with me. Before I begin the list of extraordinary contributors, I want to spotlight two of them: Jay Sanders, MD, and Scott Tibbitts.

I met Jay, the "Father of Telemedicine," in 1994 when working on my first book. He talked about discussions and research underway to meet the medical needs of people going to Mars. Through the decades, Jay and I periodically reached out to each other; he was the first person I thought of when Alastair and I decided to collaborate.

Scott Tibbitts came to me as an agent with his extraordinary story about how his little company, Starsys, made space exploration history. Scott saved Matt Damon's life . . . sort of. The space motor on the Mars Pathfinder rover that *The Martian*'s lead character used to signal to Earth that he was alive was Scott's. NASA sent it to Mars, where it arrived in January 2004. Interestingly, Scott's father Ted was the horticulturist behind the science of growing potatoes in space, so he saved Matt Damon's life, too. Scott ultimately introduced me to many people, all of whom had stories related to entrepreneurship in the space industries, and they are contributors to this book as well.

Thank you also to the following people, with whom I shared fabulous, enlightening conversations: Dr. Angel Abbud Madrid, Director, Space Resources Program, Colorado School of Mines; Dr. Theodore Tibbitts, Emeritus Professor, University of Wisconsin; Michelle Beard, Katasi; Dr. Mary Ellen Weber, STELLAR Strategies and former NASA astronaut; Dr. Robert Alan Eustace, who holds the record for the highest altitude freefall jump; Dr. Art Thompson, CEO of Sage Cheshire, Inc. and the Red Bull Stratos Technical Project, and Flight Test Museum Foundation's Chairman of the Board; Michael Dobson, long-time friend and owner of an original Apollo spacesuit; Dr. Luis Zea, Research Professor, BioServe Space Technologies, University of Colorado; Brian Sanders, co-founder and Vice President of Space Systems, Orbital Micro Systems; Blaine Pellicore, Vice President of Defense for Ursa Major Technologies; Chris McCormick, Chairman and Founder: PlanetiQ/Global Weather & Climate Solutions; Twyman Clements, founder and CEO, Space Tango; Ken Podwalski, Director Gateway at the Canadian Space Agency; Doug Howarth, founder and CEO, Multidimensional Economic

Evaluators (MEE) Inc.; and Dr. Gregory Benford, science fiction author and Professor Emeritus at the Department of Physics and Astronomy at the University of California, Irvine. If I missed anyone, please forgive me!

Thank you, as always, to Jim McCormick, who not only added to my collection of source material with important contributions, but also introduced me to Alan Eustace and Art Thompson.

The team at HarperCollins Leadership also gets a huge thank you from both of us—I know there are many of you, but here are the ones I know: Tim Burgard, Sara Kendrick, Jeff Farr, and Leigh Grossman.

Thank you also to my family and friends who let me go on and on about every big and little bit of information I learned while researching this book. And thanks to the famous and the not-so-famous entrepreneurs, scientists, and adventurers in every field who keep us looking heavenward with hope.

ENDNOTES

INTRODUCTION

1. Robert Lee Hotz, "A Hidden Hero of Apollo 11: Software," *The Wall Street Journal*, July 15, 2019, B2.

CHAPTER ONE

1. Michael Silver, "China's Dangerous Monopoly on Metals," *The Wall Street Journal*, April 16, 2019.
2. Neil DeGrasse Tyson, *Space Chronicles: Facing the Ultimate Frontier* (New York: W. W. Norton, 2012), p. 72.
3. Max Kleiber, "Animal Food for Astronauts," paper presented at the Conference on Nutrition in Space and Related Waste Problems, sponsored by the National Aeronautics and Space Administration and the National Academy of Sciences with the cooperation of the University of South Florida, Tampa, FL, April 27–30, 1964, p. 311.
4. Interview with Scott Tibbitts, November 5, 2019.
5. Theodore Tibbitts, R. Bula, R. Corey, and R. Morrow, "Cultural Systems for Growing Potatoes in Space," *Acta Horticulurae* 230 (1988): 287–9, accessed at https://www.ncbi.nlm.nih.gov/pubmed/11539774.
6. Email from Michael Dobson, December 5, 2019, based on his blog post from August 2, 2018.
7. Charles Fishman, "The Improbable Story of the Bra-maker Who Won the Right to Make Astronaut Spacesuits," *Fast Company*, July 15, 2019., accessed at https://www.fastcompany.com/90375440/the-improbable-story-of-the-bra-maker-who-won-the-right-to-make-astronaut-spacesuits.

8. Brandon Holveck, "Apollo 11 Space Mission: How Delaware Companies Made the Moon Landing Possible," *Delaware News Journal*, July 12, 2019, accessed at https://www.delawareonline.com/story/news/2019/07/12/moon-landing-how-delaware-helped-make-apollo-11-mission-possible/1684954001/.

9. Interview with Robert Alan Eustace, PhD, who made the highest-altitude skydive in history, November 14, 2019.

10. Interview with Robert Alan Eustace.

11. Interview with Robert Alan Eustace.

12. "The Stratosphere—Overview," UCAR Center for Science Education, accessed at https://scied.ucar.edu/shortcontent/stratosphere-overview.

13. Interview with Robert Alan Eustace.

14. "Suited for Space," a DuPont-Sponsored Exhibition on Extended Tour, accessed at https://www.dupont.com/products-and-services/personal-protective-equipment/articles/suited-for-space.html.

15. Tereza Pultarova, "No More Dirty Clothes! NASA Plans to Develop First Washing Machine in Space," *Space Safety Magazine*, December 2, 2011, accessed at http://www.spacesafetymagazine.com/spaceflight/life-in-orbit/dirty-clothes-nasa-plans-introduce-washing-machine-space/.

16. "Gateway Program Module(s) Continued use of NextSTEP-2 Broad Agency Announcement (BAA) Appendix A," issued by the National Aeronautics and Space Administration, Johnson Space Center Office, July 19, 2019.

17. Ioana Cozmuta, PhD, Science and Technology Corporation; Daniel Rasky, PhD, NASA ARC; Lynn Harper, NASA ARC; Robert Pittman, Lockheed-Martin, *Microgravity-Based Commercial Opportunities for Material Sciences and Life Sciences: A Silicon Valley Perspective*, Space Portal Level 2 Emerging Space Office NASA Ames Research Center, May 2014, accessed at https://www.nasa.gov/sites/default/files/ioana_hq_spaceportal_presented1.pdf.

18. Peter Rejcek, "Research in Zero Gravity: 6 Wild Projects on the International Space Station," SingularityHub, December 5, 2018, accessed at https://singularityhub.com/2018/12/05/research-in-zero-gravity-6-cool-projects-from-the-international-space-station/.

19. Peter Rejcek, "Research in Zero Gravity."

20. Luis Zea, Ph.D., ISS360: The ISS National Lab Blog, accessed at https://www.issnationallab.org/blog/fighting-cancer-with-microgravity-research/.

21. "America at the Threshold: Report of the Synthesis Group on America's Space Exploration Initiative," US Government Printing Office, Washington, D.C., 1991," p. 66.

22. Haylie Kasap, "The Race to Manufacture ZBLAN," *Upward: Magazine of the ISS National Lab*, October 19, 2019, accessed at https://upward.issnationallab.org/the-race-to-manufacture-zblan/.

23. Sarah Lewin, "Making Stuff in Space: Off-Earth Manufacturing Is Just Getting Started," *Space*, May 11, 2018, accessed at https://www.space.com/40552-space-based-manufacturing-just-getting-started.html.

24. "Ted Cruz: 'The first trillionaire will be made in space," *Politico*, June 1, 2018. Accessed at https://www.politico.com/story/2018/06/01/ted-cruz-space-first-trillionaire-616314.

25. Leonard David, "Is Asteroid Mining Possible? Study Says Yes, for $2.6 Billion," *Space Insider*, April 24, 2012, accessed at https://www.space.com/15405-asteroid-mining-feasibility-study.html.

26. "Most Valuable Asteroids in the Asteroid Belt Based on Mineral and Element Content," Statistica, last edited November 2, 2016, accessed at https://www.statista.com /statistics/656143/mineral-and-element-value-of-selected-asteroids/.

27. Leonard David, "Is Asteroid Mining Possible?"

CHAPTER TWO

1. Richard Halloran, "The Sad, Dark End of the British Empire," *Politico*, August 26, 2014, accessed at https://www.politico.com/magazine/story/2014/08/the-sad-end -of-the-british-empire-110362

2. Richard Halloran, "The Sad, Dark End of the British Empire."

3. Erin Blakemore, "How the East India Company Became the World's Most Powerful Business," *National Geographic*, September 6, 2019, accessed at https://www.national geographic.com/culture/topics/reference/british-east-india-trading-company-most -powerful-business/.

4. Erin Blakemore, "How the East India Company Became the World's Most Powerful Business."

5. Jeremy Hsu, "Don't Panic about Rare Earth Elements," *Scientific American*, May 31, 2019, accessed at https://www.scientificamerican.com/article/dont-panic-about-rare -earth-elements/.

6. Charlotte McLeod, "10 Top Countries for Rare Earth Metal Production," *Rare Earth Investing News*, May 23, 2019, accessed at https://investingnews.com/daily/resource -investing/critical-metals-investing/rare-earth-investing/rare-earth-producing -countries/.

7. James Martin and Ian Sherr, "How Apple's Daisy iPhone Recycling Robot Works," C/NET, April 18, 2019; https://www.cnet.com/news/how-apples-daisy-iphone -recycling-robot-works/.

8. "Rare Earth Recycling," Energy.gov, accessed at https://www.energy.gov/science/bes /articles/rare-earth-recycling.

9. Iris Bruijn, *Ships Surgeons of the Dutch East India Company: Commerce and the Progress of Medicine in the Eighteen Century* (Rapenburg, Netherlands: Leiden University Press, 2009), p. 15.

10. Maryann Karinch, *Telemedicine: What the Future Holds When You're Ill* (New Horizon Press, 1994), Chapter 4.

11. Bernard Harris, MD, "Space Adaptation Syndrome and the Implications of Telemedicine," Presented at the May Telemedicine Symposium, October 1993.

12. Bernard Harris, "Space Adaptation Syndrome and the Implications of Telemedicine."

13. Interview with Jay Sanders, MD, "The Father of Telemedicine," October 19, 2019.

14. "Attached," *Star Trek: The Next Generation*, original airdate November 8, 1993; written by Nick Sagan.

CHAPTER THREE

1. Interview with Scott Tibbitts.

2. Interview with Angel Abbud-Madrid, PhD, Director, Center for Space Resources of Colorado School of Mines, October 30, 2019.

3. "List of small business grants in the UK," Entrepreneur Handbook, September 19, 2019, accessed at https://entrepreneurhandbook.co.uk/grants-loans/.

4.   The National Academies Press, SBIR at NASA (2016), Chapter: 4 SBIR Awards, accessed at https://www.nap.edu/read/21797/chapter/6.

5.   SBIR × STTR Award Information. Accessed at https://www.sbir.gov/sbirsearch /award/all?page=17&f%5B0%5D=im_field_agencies%3A105737&f%5B1%5D =itm_field_award_yr%3A2018&solrsort=fts_field_award_amt%20desc.

6.   Scott Tibbitts, "Entrepreneurial Space Companies: Lessons Learned from Those That Have Survived and Thrived," remarks prepared for and presented during the 59th annual Aerospace Conference in Glasgow, Scotland, 2008.

7.   Interview with Chris McCormick, founder of Broad Reach, now Chairman and Co-founder of PlanetiQ, November 7, 2018.

8.   Interview with Chris McCormick.

9.   Scott Tibbitts, "Entrepreneurial Space Companies: Lessons Learned from Those That Have Survived and Thrived."

10. https://www.spacebandits.io/startups.

11. Interview with Scott Tibbitts.

12. From the as-yet unpublished work by Scott Tibbitts, tentatively titled *From the Garage to Mars.*

13. Archived footage of the Mars Exploration Rover's New Briefing, NASA's Jet Propulsion Laboratory, California Institute of Technology, 9:30 PM PST, January 3, 2004.

14. Archived footage of the Mars Exploration Rover's New Briefing, January 3, 2004.

15. Scott Tibbitts, *From the Garage to Mars.*

16. Interview with Eric Anderson, President and COO, SEAKR Engineering, November 14, 2019.

17. *PC Magazine*'s encyclopedia defines bubble memory as "an early non-volatile magnetic storage device...bubble memory was about as fast as a slow hard disk but it held its content without power. As hard disks greatly improved in the 1980s, bubble memory was abandoned even though it was well suited for rugged applications," accessed at https://www.pcmag.com/encyclopedia/term/39006/bubble-memory

18. Interview with Eric Anderson.

19. Interview with Eric Anderson.

20. Interview with Brian Sanders, co-founder and Vice President of Space Systems, Orbital Micro Systems, November 15, 2019.

21. Landsat Science, accessed at https://landsat.gsfc.nasa.gov/

22. Interview with Brian Sanders.

23. Interview with Brian Sanders.

24. Interview with Chris McCormick.

25. Pamela G. Hollie, "Union Oil Successes Scarcer in Indonesia," *The New York Times*, January 19, 1981.

CHAPTER FOUR

1.   "Where Today Are the Apollo 17 Goodwill Moon Rocks?" CollectSpace.com, accessed at http://www.collectspace.com/resources/moonrocks_goodwill.html.

2.   Alexander Hamilton, General Introduction to what became known as *The Federalist Papers*, initially published anonymously in the *Independent Journal*, October 1787, accessed at https://www.congress.gov/resources/display/content/The+Federalist +Papers#TheFederalistPapers-1.

3.   "Treaty on Principles Governing the Activities of States in the Exploration and Use of Outer Space, Including the Moon and Other Celestial Bodies," *United Nations Treaties*

and Principles on Outer Space (New York: United Nations, 2002), p. 11, accessed at http://www.unoosa.org/pdf/publications/STSPACE11E.pdf.

4. "Title IV—Space Resource Commercial Exploration And Utilization," US Commercial Space Launch Competitiveness Act," PUBLIC LAW 114–90—NOV. 25, 2015, § 51302, accessed at https://www.congress.gov/114/plaws/publ90/PLAW-114publ90.pdf.

5. D. A. Broniatowski, G. Ryan Faith, and Vincent G. Sabathier, "The Case for Managed International Cooperation in Space Exploration," International Space Exploration Update, Center for Strategic and International Studies, Washington, D.C., 2006, accessed at https://web.mit.edu/adamross/www/BRONIATOWSKI_ISU07.pdf.

6. Nathan C. Goldman, *American Space Law: International and Domestic* (Ames: Iowa State University Press, 1988), p. vii.

7. Nathan C. Goldman, *American Space Law*, p. vii.

8. "The Missile Technology Control Regime at a Glance," Arms Control Association, July 2017, accessed at https://www.armscontrol.org/factsheets/mtcr.

9. Debra Werner, "The Torture of CFIUS," *Space News*, October 7, 2019, pp. 16–17.

10. Amy Thompson, "Traffic Jams from Satellite Fleets Are Imminent—What It Means for Earth," *Observer*, September 5, 2019, accessed at https://observer.com/2019/09/satellite-space-congestion-spacex-starlink-esa-aeolus/#:~:text=Aeolus%20launched%20in%20August%202018,in%20May%20of%20this%20year.

11. See https://www.iso.org/ics/49.140/x/.

12. Sandra Erwin, "U.S. Space Command Eager to Hand Over Space Traffic Duties to Commerce Department," *Space News*, November 17, 2019, accessed at https://spacenews.com/u-s-space-command-eager-to-hand-over-space-traffic-duties-to-commerce-department/.

13. Madhu Thangavelu, "A U.S. Department of Space?" *Space News*, July 4, 2012, accessed at https://spacenews.com/us-department-space/.

CHAPTER FIVE

1. Avni Shah, "Space or Nothing: How USC Sent the First Student-Built Rocket to Outer Space," *USC Viterbi*, Fall 2019, p. 43.

2. Gwynne Shotwell, "SpaceX's Plan to Fly You Across the Globe in 30 Minutes," TED Talk, April 2018, accessed at https://www.ted.com/talks/gwynne_shotwell_spacex_s_plan_to_fly_you_across_the_globe_in_30_minutes?language=en.

3. "*Chandrayaan-2: Failure is part of the 'Big Game,' Shouldn't Discourage the Science Community, Says NASA-JPT CTO,*" *Technology News, December 16, 2019*, accessed at https://www.firstpost.com/tech/science/chandrayaan-2-failure-is-part-of-the-big-game-shouldnt-discourage-the-science-community-says-nasa-jpl-cto-7788351.html.

4. Interview with Blaine Pellicore, Vice President of Defense for Ursa Major Technologies, November 12, 2019.

5. Christian Davenport, "After 2016 Rocket Explosion, Elon Musk's SpaceX Looked Seriously at Sabotage," *The Washington Post*, February 26, 2018, accessed at https://www.washingtonpost.com/news/the-switch/wp/2018/02/26/after-2016-rocket-explosion-elon-musks-spacex-looked-seriously-at-sabotage/.

6. Interview with Blaine Pellicore.

7. Brian Fishbine, Robert Hanrahan, Steven Howe, Richard Malenfant, Carolynn Scherer, Haskell Sheinberg, and Octavio Ramos Jr., "Nuclear Rockets: To Mars and Beyond," *National Security Science* I (2011), Los Alamos National Laboratory, accessed at https://www.lanl.gov/science/NSS/issues/NSS-Issue1-2011.pdf.

8. Interview with Blaine Pellicore.
9. Kaitlyn Johnson, "What Does the Trump Administration's New Memorandum Mean for Nuclear-Powered Space Missions?" Center for Strategic & International Studies, August 28, 2019, accessed at https://www.csis.org/analysis/what-does-trump -administrations-new-memorandum-mean-nuclear-powered-space-missions.
10. Jeff Foust, "'What the Hell Happened?': The Rise and Fall of Suborbital Space Tourism Companies," *Space News*, June 13, 2017, accessed at https://spacenews.com/what -the-hell-happened-the-rise-and-fall-of-suborbital-space-tourism-companies/.
11. See Reaction Engines homepage at https://www.reactionengines.co.uk/.
12. Jon Kelvey, "Into the Future at 6,000 MPH," *Air & Space*, January 2020, p. 31.
13. Jon Kelvey, "Into the Future at 6,000 MPH," p. 33.
14. Jon Kelvey, "Into the Future at 6,000 MPH," p. 31.
15. Robotics Systems, Jet Propulsion Laboratory, California Institute of Technology, accessed at https://www-robotics.jpl.nasa.gov/systems/system.cfm?System=11.
16. Interview with Don Pickering, CEO, Olis Robotics, December 2, 2019.
17. Interview with Don Pickering.
18. Mike Wall, "President Obama's Space Legacy: Mars, Private Spaceflight and More," Space.com, January 20, 2017, accessed at https://www.space.com/35394-president -obama-spaceflight-exploration-legacy.html.

CHAPTER SIX

1. Interview with Luis Zea, PhD, Assistant Research Professor—Bionautics at BioServe Space Technologies, a research institute within the University of Colorado in Boulder, Colorado, October 13, 2019.
2. Interview with Twyman Clements, co-founder and CEO of Space Tango, October 28, 2019.
3. "Budweiser Takes Next Step to Be the First Beer on Mars," Anheuser-Busch, November 21, 2017, accessed at https://www.anheuser-busch.com/newsroom/2017/11/budweiser -takes-next-step-to-be-the-first-beer-on-mars.html.
4. Interview with Twyman Clements.
5. Twyman Clements, as quoted in "Space Tango Announces Launch of LambdaVision Retinal Implant Manufacturing Payload on SpaceX CRS-16," December 5, 2018, accessed at https://spacetango.com/space-tango-announces-launch-of-lambdavision -retinal-implant-manufacturing-payload-on-spacex-crs-16/.
6. Interview with Twyman Clements.
7. Interview with Twyman Clements.
8. Interview with Luis Zea.
9. Interview with Luis Zea.
10. Interview with Luis Zea.
11. Interview with Ken Podwalsky, Director Gateway, Canadian Space Agency, November 6, 2019.
12. Interview with Ken Podwalsky.
13. Prime Minister Justin Trudeau, tweeting on February 20, 2019, as quoted on Space .com, March 1, 2019, accessed at https://www.space.com/nasa-lunar-gateway-canada -canadarm3-robot-arm.html.
14. See, for instance, Susanne Barton and Hannah Recht, "The Massive Prize Luring Miners to the Stars," Bloomberg, March 8, 2018, accessed at https://www.bloomberg.com /graphics/2018-asteroid-mining/.

15. Interview with Angel Abbud-Madrid.
16. Jacob Gershman, "Who Owns the Moon?," *The Wall Street Journal* Supplement "Apollo 11|50 Years Later," July 15, 2019, R4.
17. Korey Hanes, "Japanese Asteroid Mission Touches Down on Ryugu, Collects Sample," Astronomy.com, July 11, 2019, accessed at http://www.astronomy.com/news/2019 /07/japanese-asteroid-mission-touches-down-on-ryugu-collects-sample.
18. Interview with Angel Abbud-Madrid.
19. Interview with Angel Abbud-Madrid.
20. Interview with Angel Abbud-Madrid.
21. A. A. Kamel, Ph.D. dissertation, Stanford University, 1969.

CHAPTER SEVEN

1. Gerard K. O'Neill on Habitat Design, Space Studies Institute archives, 1981, accessed at https://www.youtube.com/watch?v=eDS42C32xTU
2. Kyla Edison, Geology & Material Science Technical for the Pacific International Space Center for Exploration Systems (PISCES), "Homesteading New Worlds," *ad Astra*, 2019-3, p. 44.
3. Kyla Edison, "Homesteading New Worlds."
4. Kyla Edison, "Homesteading New Worlds," p. 47.
5. Fred Scharmen, "Jeff Bezos Dreams of a 1970s Future," *CityLab*, May 13, 2019, accessed at https://www.citylab.com/perspective/2019/05/space-colony-design-jeff -bezos-blue-origin-oneill-colonies/589294/.
6. Michelle A. Ricker and Shelby Thompson, "Developing a Habitat for Long Duration, Deep Space Missions," NASA Technical Reports, FLEX-2012.05.3.8x12222, accessed at https://ntrs.nasa.gov/archive/nasa/casi.ntrs.nasa.gov/20120008183.pdf.
7. Elizabeth Howell, "NASA Will Test 5 Habitat Designs for Its Lunar Gateway Space Station," Space.com, April 5, 2019. Accessed at https://www.space.com/nasa-lunar -gateway-habitat-designs-testing.html.
8. "NASA Selects Two New Space Tech Research Institutes for Smart Habitats," April 8, 2019, accessed at https://www.nasa.gov/press-release/nasa-selects-two-new-space-tech -research-institutes-for-smart-habitats.
9. "NASA Selects Two New Space Tech Research Institutes for Smart Habitats."
10. Sherry Stokes, "A Smarter Habitat for Deep Space Exploration," Phys.org, September 25, 2019, accessed at https://phys.org/news/2019-09-smarter-habitat-deep-space -exploration.html.
11. "Exploring Life-Enhancing Built Environments," USC Viterbi School of Engineering Newsletter, Fall 2019.

CHAPTER EIGHT

1. Adam Rogers, "Space Tourism Isn't Worth Dying For," *Wired*, October 31, 2014, accessed at https://www.wired.com/2014/10/virgin-galactic-boondoggle/.
2. See https://wttc.org/.
3. Glenn Research Center, NASA.gov; https://www.nasa.gov/centers/glenn/events/tour _zgf.html; https://www.gozerog.com/index.cfm?fuseaction=Experience.welcome
4. See https://www.gozerog.com/.
5. "Roscosmos and Space Adventures Sign Contract for Orbital Space Tourist Flight," Spaceadventures.com, press release, February 19, 2019, accessed at https://space

adventures.com/roscosmos-and-space-adventures-sign-contrat-for-orbital-space
-tourist-flight/.

6. "The First Ever Space Hotel—Galactic Suite the Outer Space Resort," Design Build
    Network, accessed at https://www.designbuild-network.com/projects/galactic-suite/.

7. Emily Clark, "Holiday in Orbit: Glactic Suite Space Resort Opening in 2012," New At-
    las, August 22, 2007, accessed at https://newatlas.com/holiday-in-orbit-galactic-suite
    -space-resort-opening-2012/7811/.

8. Jeff Foust, "'What the Hell Happened?'"

9. Interview with Art Thompson, co-founder and Vice President of Sage Cheshire Aero-
    space, November 20, 2019.

10. Samantha Mathewson, "XCOR 'Space Tourists' Push for Ticket Refunds: Report,"
    Space.com, January 5, 2019, accessed at https://www.space.com/42912-xcor
    -aerospace-ticket-holders-request-refund.html.

11. Samantha Mathewson, "XCOR 'Space Tourists' Push for Ticket Refunds: Report."

12. Samantha Mathewson, "XCOR 'Space Tourists' Push for Ticket Refunds: Report."

13. Interview with Art Thompson.

14. Interview with Art Thompson.

15. The Golden Spike Company media kit, accessed at https://media.boingboing.net
    /wp-content/uploads/2012/12/GSC-Officers_Advisors_Partners.pdf.

16. Ruqayyah Moynihan and Thomas Giraudet, "Richard Branson Wants Virgin Galactic
    to Send People to Space Every 32 Hours by 2023," *Business Insider*, September 10,
    2019, accessed at https://www.businessinsider.com/branson-virgin-galactic-people
    -space-every-32-hours-2019-9.

17. Michael Goldstein, "Virgin Galactic: From Space to the Stock Market," *Forbes*, October
    30, 2019, accessed at https://www.forbes.com/sites/michaelgoldstein/2019/10/30
    /virgin-galactic-from-space-to-the-stock-market/#604cf7297b08.

18. Zhao Lei, "Space Tourism Just Around the Corner, Rocket Maker Says," *The Telegraph*,
    July 25, 2018, accessed at https://www.telegraph.co.uk/china-watch/technology
    /space-tourism/.

19. "Space Tourism: 5 Companies That Will Make You an Astronaut," Revfine, accessed at
    https://www.revfine.com/space-tourism/; https://www.boeing.com/space/starliner/.

20. See https://www.boeing.com/space/starliner/.

21. Duncan Madden, "Mankind's First Space Hotel Is Coming in 2021—Probably," *Forbes*,
    March 9, 2018, accessed at https://www.forbes.com/sites/duncanmadden/2018
    /03/09/mankinds-first-space-hotel-is-coming-in-2021-probably/#783988195bed.

22. "Space Tourism: 5 Companies That Will Make You An Astronaut."

23. Title 14: Aeronautics and Space; Part 460 Human Space Flight Requirements; Subpart
    B—Launch and Reentry with a Space Flight Participant.

CHAPTER NINE

1. Robert Zubrin, *The Case for Space* (Amherst, NY: Prometheus Books, 2019), p. 101.

2. Jim Bridenstine, NASA Administrator, in an interview with C-SPAN, July 12, 2019,
    accessed at https://www.c-span.org/video/?462496-1/newsmakers-jim-bridenstine.

3. Jim Bridenstine, NASA Administrator in an exchange recorded by C-SPAN, July 19,
    2019, accessed at https://www.c-span.org/video/?462853-1/president-trump-meets
    -apollo-11-astronauts.

4. Mike Wall, "Bill Nye: It's Space Settlement, Not Colonization," Space.com, October
    23, 2019, accessed at https://www.space.com/bill-nye-space-settlement-not-colonization
    .html?utm_source=notification.

5.   Interview with Angel Abbud-Madrid.

6.   Joel Anderson, "Here's How Much It Would Really Cost to Build a Moon Colony,"
     Yahoo! Finance, September 13, 2019, accessed at https://finance.yahoo.com/news
     /much-really-cost-build-moon-184949330.html

7.   Leonard David, *Moon Rush: The New Space Race* (Washington, D.C.: National Geo-
     graphic, 2019), p. 96.

8.   Interview with Luis Zea.

9.   "Technology Readiness Level (TRL)," AcqNotes, accessed at http://acqnotes.com
     /acqnote/tasks/technology-readiness-level.

10. Ioana Cozmuta, PhD, Science and Technology Corporation; Daniel Rasky, PhD,
     NASA ARC; Lynn Harper, NASA ARC; Robert Pittman, Lockheed-Martin,
     *Microgravity-Based Commercial Opportunities for Material Sciences and Life Sciences: A
     Silicon Valley Perspective*, Space Portal Level 2 Emerging Space Office NASA Ames
     Research Center, May, 2014, accessed at https://www.nasa.gov/sites/default/files
     /ioana_hq_spaceportal_presented1.pdf.

11. Cozmuta et al, *Microgravity-Based Commercial Opportunities for Material Sciences and
     Life Sciences*, p. 5.

12. Cozmuta et al, *Microgravity-Based Commercial Opportunities for Material Sciences and
     Life Sciences*.

13. "Aretech's ZeroG Used in Research to Help Restore the Ability to Walk Following
     Spinal Cord Injury," *Neuroscience News*, September 24, 2015, accessed at https://www
     .aretechllc.com/2015/09/zerog-used-in-research-to-help-restore-the-ability-to-walk
     -following-spinal-cord-injury/.

CHAPTER TEN

1.   Joel Levine, PhD, "The Exploration and Colonization of Mars: Why Mars? Why Hu-
     mans?" TEDxRVA, July 31, 2015, accessed at https://www.youtube.com/watch?v
     =YzhSmnGcSkE.

2.   Joel Levine, "The Exploration and Colonization of Mars: Why Mars? Why Humans?"

3.   Bas Lansdorp at TEDxDelft, "Getting Humanity to Mars," November 5, 2012. Ac-
     cessed at https://www.youtube.com/watch?v=qnWcYnyvkBo.

4.   Rebecca Boyle, "Elon Musk's Boldest Announcement Yet: In a Gorgeous New Video,
     the SpaceX CEO Lays Out His Vision for a Human Civilization on Mars," *The Atlan-
     tic*, September 27, 2017, accessed at https://www.theatlantic.com/science/archive
     /2016/09/elon-musk-mars/501794/.

5.   Boeing CEO Dennis Muilenburg in an interview with Jim Cramer, *Mad Money with
     Jim Cramer*, CNBC, December 7, 2017, accessed at https://www.cnbc.com/video
     /2017/12/07/boeing-ceo-talks-trump-taxes-tesla-growth-and-more.html.

6.   Bas Lansdorp at TEDxDelft, "Getting Humanity to Mars."

7.   Sarah Baatout, PhD, Director of the Radiation Unit, Belgian Nuclear Research Center,
     "How Can We Better Protect Astronauts From Space Radiation?" TEDxAntwerp,
     March 11, 2019, accessed at https://www.youtube.com/watch?v=e607t5AeIxk.

8.   Sarah Baatout, "How Can We Better Protect Astronauts From Space Radiation?"

9.   Matt Williams, "How Bad Is the Radiation on Mars?" Phys.Org, November 21, 2016,
     accessed at https://phys.org/news/2016-11-bad-mars.html.

10. "Real Martians: How to Protect Astronauts from Space Radiation on Mars," Moon
     to Mars, NASA.gov, September 30, 2015, accessed at https://www.nasa.gov
     /feature/goddard/real-martians-how-to-protect-astronauts-from-space-radiation
     -on-mars

11. Susie Neilson, "Mold from Chernobyl seems to feed on radiation, and new research suggests it could help protect astronauts in space," *Business Insider*, July 28, 2020, accessed at https://www.businessinsider.com/chernobyl-mold-protect-astronauts-from -radiation-in-space-2020-7.

12. Sarah Baatout, "How Can We Better Protect Astronauts From Space Radiation?"

13. Interview with Jay Sanders.

14. B. Yang, X. Fang, and J. Kong, "In Situ Sampling and Monitoring Cell-Free DNA of the Epstein-Barr Virus from Dermal Interstitial Fluid Using Wearable Microneedle Patches," *ACS Applied Materials and Interfaces* 11 no. 42 (2019): 38448–38458. doi: 10.1021/acsami.9b12244.

15. Interview with Jay Sanders.

16. Joel Levine, "The Exploration and Colonization of Mars: Why Mars? Why Humans?"

CONCLUSION

1. Interview with Angel Abbud-Madrid.

2. Jean Wright as quoted by Francis French, "A Suit for the Shuttle," *ad Astra*, 2019-3, p. 33.

3. Francis French, "A Suit for the Shuttle," p. 33.

INDEX

Note: Page numbers in *italics* refer to photos, charts, or tables.

ABOUT THE AUTHORS

 ALASTAIR STORM BROWNE has been a life-long space advocate and holds an MS degree in space studies from the University of North Dakota. This is his first book published, a product of his life's work, which he began in 1990 after he received his degree. His hobbies include writing, traveling, and playing guitar with friends. He is at present working on a new book about rebuilding the United States. Alastair lives in Durham, North Carolina.

 MARYANN KARINCH is the author of thirty-two books, including *Nothing But the Truth* and *Lessons from the Edge*. She is the founder of The Rudy Agency, a full-service literary agency based in Estes Park, Colorado (www .RudyAgency.com).